AF827386

Atlas of the Deep-Water Seabed

Boris Dorschel · Andrew J. Wheeler
Xavier Monteys · Koen Verbruggen

Atlas of the Deep-Water Seabed

Ireland

Springer

Boris Dorschel
Department of Geology and Environmental Research Institute
University College Cork
Ireland
b.dorschel@ucc.ie

Andrew J. Wheeler
Department of Geology and Environmental Research Institute
University College Cork
Ireland
a.wheeler@ucc.ie

Xavier Monteys
Geological Survey of Ireland
Dublin, Ireland
Xavier.Monteys@gsi.ie

Koen Verbruggen
Geological Survey of Ireland
Dublin, Ireland
Koen.Verbruggen@gsi.ie

ISBN 978-90-481-9375-2 e-ISBN 978-90-481-9376-9
DOI 10.1007/978-90-481-9376-9
Springer Dordrecht Heidelberg London New York

Library of Congress Control Number: 2010934111

© Springer Science+Business Media B.V. 2010
No part of this work may be reproduced, stored in a retrieval system, or transmitted in any form or by any means, electronic, mechanical, photocopying, microfilming, recording or otherwise, without written permission from the Publisher, with the exception of any material supplied specifically for the purpose of being entered and executed on a computer system, for exclusive use by the purchaser of the work.

Cover design by deblik; seabed photos by MARUM – Center for Marine Environmental Sciences, University of Bremen, Alfred-Wegener-Institut für Polar- und Meeresforschung, Institut Français de Recherche pour l'Exploitation de la Mer, Department of the Environment, Heritage and Local Government, the Marine Institute and the Geological Survey of Ireland; map by X. Monteys.

Printed on acid-free paper

Springer is part of Springer Science+Business Media (www.springer.com)

This atlas is dedicated to

*Kolja, Zoë, Malin and Katja Dorschel, Nessa and Malachy Wheeler,
Moira Ní Loingsigh, Dolors, Silvia and Manel Monteys, Bernie, Jana,
Katie and Conor Verbruggen*

Can ye fathom the ocean, dark and deep, where the mighty waves and the grandeur sweep?

Frances Jane Crosby
24 March 1820 – 12 February 1915

Preface

The inspiration for this atlas came from a project locating all cold-water coral reefs and associated carbonate mounds on the Irish seabed. The fact that coral reefs even exist on Ireland's deep-water seabed may come as a surprise to some readers but in fact this is an area of the world's oceans where these cold-water features arguably reach their most impressive proportions and densities.

The detection of potential reefs only a few hundred metres wide required high resolution bathymetric (or seabed topographic) data. Fortunately, the Irish National Seabed Survey (INSS) has collected just such a data set over the course of a 6 year period starting in 1999. This represents an impressive and comprehensive mapping of the entire Irish deep-water seabed (deeper than 200 m) covering over half a million square kilometres and extending up to nearly a thousand kilometres off the west coast of Ireland. Using the INSS data, and funded by its successor INFOMAR, we were able to investigate a coherent 414,000 km^2 area of the world's deep ocean seabed at a level of detail that has never been achieved before.

As we systematically scoured the maps for coral reefs and mounds, we discovered a plethora of other submarine features imaged in startling detail which were a wonder to see. We realised that the initially planned atlas of reef features represented just one aspect of what was of interest and felt somewhat obliged to "lift the curtain" on this deep and dark ocean realm.

We hope you will share our excitement in exploring Ireland's deep-water seabed and maybe even become inspired to find out more about this hidden world several hundreds of metres below the sea surface. Our aim is to demonstrate that the oceans are not simply deep water-filled basins but contain fascinating "land-scape" structures. These landscape features, such as submarine canyons, seamounts, escarpments and mounds, are also home to a diverse array of complex habitats. We want to demonstrate the amazing variety and beauty of life on the deep-sea seabed. Our goal is to invite the reader on a voyage to the largely unexplored canyons, sub-sea mountains and cliffs of the Irish underwater realm.

Enjoy the trip!

The Authors

Acknowledgements

We wish to express our sincere gratitude to those who made this atlas possible in the first place, namely the many people who worked together to acquire, process and manage the marine bathymetric dataset as the Irish National Seabed Survey, carried out under the Director of the Geological Survey of Ireland, Dr. Peadar McArdle. Furthermore, we are grateful to Moira Ní Loingsigh for her comments on the text and Glenn Nolan for his help with the oceanographic data sets. We also thank Eamonn Kelly, Thomas Furey, Doug Masson, Volker Ratmeyer and Peter Croker for providing imagery.

Contents

Contributors

Dr. Boris Dorschel

Dr Boris Dorschel is a research scientist at University College Cork, Ireland. He graduated in geology and completed his Ph.D. thesis on cold-water coral carbonate mounds at the University of Bremen and MARUM – Center for Marine Environmental Sciences. Since then, he has been working in collaborative research projects on the influence of environmental factors on carbonate mound development.

Dr. Andy Wheeler

Dr. Andy Wheeler studied geology at University of Liverpool and completed a Ph.D. at the University of Cambridge in 1994. He is currently Senior Lecturer in the Department of Geology and Vice Head of the School of Biological, Earth and Environmental Sciences at University College Cork, Ireland. He has published extensively on aspects of deep-sea marine geology including coral carbonate mounds, seabed mapping and seabed sedimentary processes.

Xavier Monteys

Xavier Monteys is a marine geologist with the Geological Survey of Ireland (GSI) since 2001. He graduated at the University Autònoma of Barcelona (UAB) and undertook postgraduate studies at the University of Maynooth (Ireland). He was involved in numerous deep-water mapping campaigns as part of the Irish National Seabed Survey (INSS) and is currently engaged in the INFOMAR project mapping the shallow environments.

Koenraad (Koen) Verbruggen

Koenraad (Koen) Verbruggen is a Principal Geologist & Section Head with the Geological Survey of Ireland (GSI). He graduated in Geology followed by a postgraduate in Petroleum Geology, both at University College Dublin and after a career in the resource industry internationally, joined the GSI in 2000. Since then, one of his main tasks has been managing the Project INFOMAR (Integrated Mapping for Sustainable Development of Irelands Marine Resource) the successor to INSS.

Part I

Background & Overview

1 Introduction

B. Dorschel, A.J. Wheeler, X. Monteys, and K. Verbruggen

Humans' perception of their surroundings is dominantly sensed through the medium of sight. What is hidden from the eye most often does not receive our attention and gets ignored. As Ireland's deep-water seabed is hidden under several hundreds of metres of water, it has also historically received limited attention: what is out of sight is so often out of mind. This fundamentally changed when Ireland realised the scientific, as well as commercial, importance of its underwater realm.

Just to put things into perspective: Ireland and its sub-sea territories cover an area of more than half a million square kilometres, more than the size of mainland France. An area like this is equal to a square with each side more than 800 km long or the size of more than 94 million rugby pitches – in other words: it is pretty large. Sub-sea Ireland is in fact 9.1 times the size of the Republic of Ireland and stretches from the shallow Irish Sea between Britain and Ireland far into the northeast Atlantic to the margin of the Iceland Basin, nearly 1,000 km west of Connemara's coast. Away from the coast, the Irish seabed drops gradually down to almost 5,000 m below sea level (bsl) in the Porcupine Abyssal Plain and the Iceland Basin (Fig. 1.1).

Due to technical limitations, up until very recently, human seabed activities on the seabed such as fishing and exploration were mostly restricted to the shallow shelves (<200 m bsl). Furthermore, topographic features on the seabed are not considered shipping hazards if they do not reach up to near the sea surface. As a consequence, the deep seabed received very little attention and until very recently, only very crude maps with kilometre-size resolution existed for these vast areas; most of these are based on Victorian British Navy depth soundings. In the latter part of the last century, sonar images were produced for some areas but they often required expert interpretation.

Only in recent years, when deep sea fishing, telecommunication and exploration industries moved into areas as deep as 1,600 m bsl, did the increasing commercial interest demand high resolution bathymetric information. For a long time, however, a high resolution map of the deep seabed remained unrealisable. For example, even the positioning of ships was not accurate enough for acquiring high resolution geographic data.

B. Dorschel (✉) and A.J. Wheeler
Department of Geology & Environmental Research Institute, University College Cork, Cork, Ireland
e-mails: b.dorschel@ucc.ie; a.wheeler@ucc.ie

X. Monteys and K. Verbruggen
Geological Survey of Ireland, Beggars Bush, Haddington Road, Dublin 4, Ireland
e-mails: Xavier.Monteys@gsi.ie; Koen.Verbruggen@gsi.ie

B. Dorschel et al., *Atlas of the Deep-Water Seabed: Ireland*,
DOI 10.1007/978-90-481-9376-9_1, © Springer Science+Business Media B.V. 2010

Fig. 1.1 **Ireland's designated continental shelf.** The Irish designated continental shelf (highlighted in yellow) represents the part of the Atlantic seabed that is connected to mainland Ireland by continental crust (see also the chapter 1.1). The exact limits of some of these areas are still subject to international negotiations with neighbouring countries. Globe image copyright © 2010 ESRI

Rapid advances in ocean mapping technologies over the past decades eventually made it possible to take on the task of mapping the deep seabed in high resolution. The development of the multibeam echosounder, and very accurate satellite-based global positioning systems (GPS) in particular, provided the tools necessary to map extensive areas of the seabed at a resolution sharp enough to resolve key features (Fig. 1.3). With these new tools at hand, a full coverage, high resolution bathymetric mapping of the Irish seabed became feasible. It is the results of this mapping which we wish to showcase here.

This atlas has been divided into four main sections:

SECTION I: BACKGROUND & OVERVIEW (Chapters 1 to 6) is an introduction to the Atlas with information on the Irish seabed, bathymetric data sets and seabed features.

SECTION II: THEMATIC ATLAS (Chapters 7 to 11) showcases the different seabed features and supplies more specific background information.

SECTION III: REGIONAL ATLAS (Chapters 12 to 18) places all features in their correct spatial context and is divided into seven areas reflecting the main banks, troughs and basins that structure the Irish seabed.

SECTION IV: GOING DEEPER... (Chapter 19) is a guide to the data sets behind the atlas allowing the readers to go on exploring on their own.

1.1 How far Ireland reaches and why?

Drawing boundaries is a tricky business and the Article 76 of the United Nations Convention on the Law of the Sea sets out the codes and guidelines for defining the legal limit to the national underwater seabed (Fig. 1.2). As you can see, it is quite complex to legally put a simple boundary on a map. The extension of the Irish seabed has however implications on the exploration of resources and is therefore of socio-economic importance.

Article 76 of the United Nations Convention on the Law of the Sea

1. The continental shelf of a coastal State comprises the seabed and subsoil of the submarine areas that extend beyond its territorial sea throughout the natural prolongation of its land territory to the outer edge of the continental margin, or to a distance of 200 nautical miles from the baselines from which the breadth of the territorial sea is measured where the outer edge of the continental margin does not extend up to that distance.
2. The continental shelf of a coastal State shall not extend beyond the limits provided for in paragraphs 4 to 6.
3. The continental margin comprises the submerged prolongation of the land mass of the coastal State, and consists of the seabed and subsoil of the shelf, the slope and the rise. It does not include the deep ocean floor with its oceanic ridges or the subsoil thereof.
4. (a) For the purposes of this Convention, the coastal State shall establish the outer edge of the continental margin wherever the margin extends beyond 200 nautical miles from the baselines from which the breadth of the territorial sea is measured, by either:
   (i) a line delineated in accordance with paragraph 7 by reference to the outermost fixed points at each of which the thickness of sedimentary rocks is at least 1 per cent of the shortest distance from such point to the foot of the continental slope; or
   (ii) a line delineated in accordance with paragraph 7 by reference to fixed points not more than 60 nautical miles from the foot of the continental slope.
   (b) In the absence of evidence to the contrary, the foot of the continental slope shall be determined as the point of maximum change in the gradient at its base.
5. The fixed points comprising the line of the outer limits of the continental shelf on the seabed, drawn in accordance with paragraph 4 (a)(i) and (ii), either shall not exceed 350 nautical miles from the baselines from which the breadth of the territorial sea is measured or shall not exceed 100 nautical miles from the 2,500 metre isobath, which is a line connecting the depth of 2,500 metres.
6. Notwithstanding the provisions of paragraph 5, on submarine ridges, the outer limit of the continental shelf shall not exceed 350 nautical miles from the baselines from which the breadth of the territorial sea is measured. This paragraph does not apply to submarine elevations that are natural components of the continental margin, such as its plateau, rises, caps, banks and spurs.
7. The coastal State shall delineate the outer limits of its continental shelf, where that shelf extends beyond 200 nautical miles from the baselines from which the breadth of the territorial sea is measured, by straight

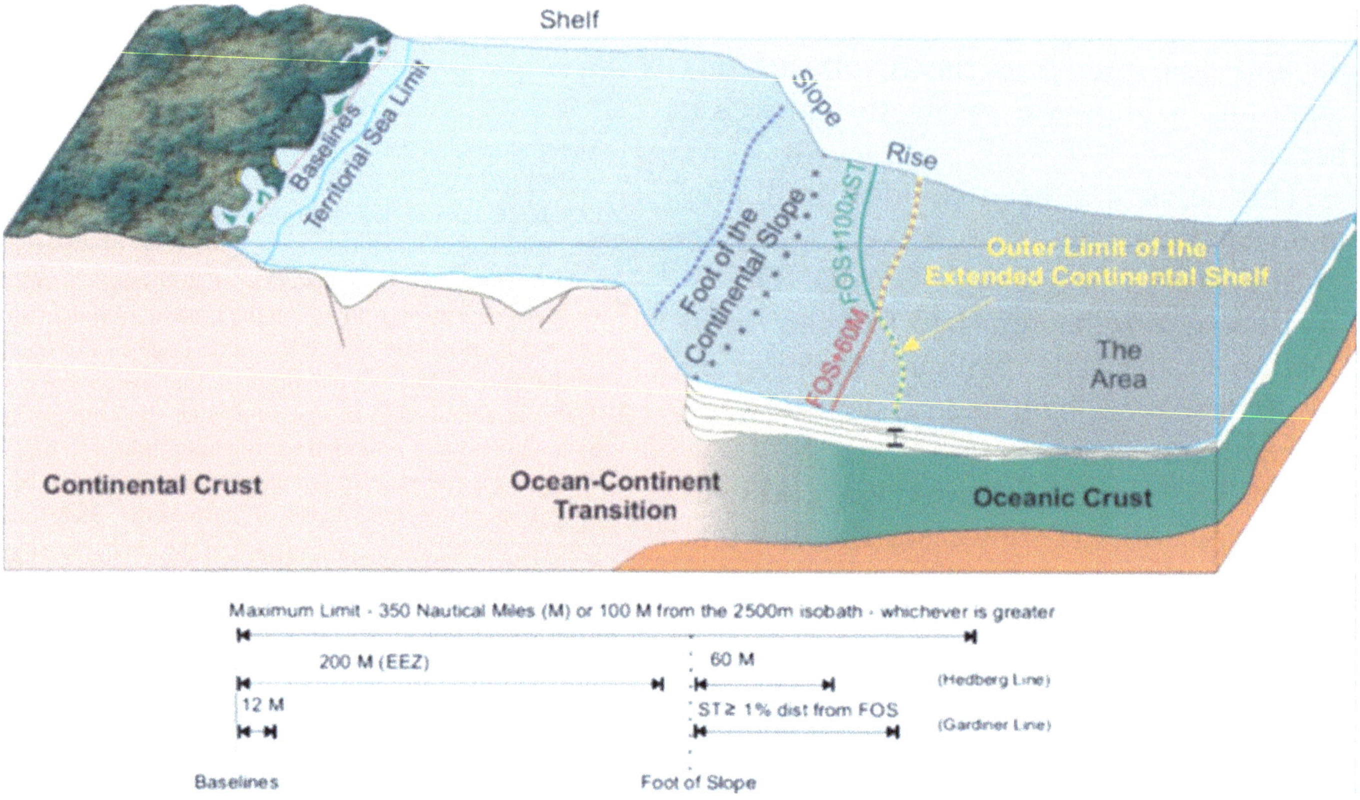

Fig. 1.2 **Delineation of the Irish Continental Shelf.** Block diagram prepared by Petroleum Affairs Division/DCENR technical staff

lines not exceeding 60 nautical miles in length, connecting fixed points, defined by coordinates of latitude and longitude.

8. Information on the limits of the continental shelf beyond 200 nautical miles from the baselines from which the breadth of the territorial sea is measured shall be submitted by the coastal State to the Commission on the Limits of the Continental Shelf set up under Annex II on the basis of equitable geographical representation. The Commission shall make recommendations to coastal States on matters related to the establishment of the outer limits of their continental shelf. The limits of the shelf established by a coastal State on the basis of these recommendations shall be final and binding.
9. The coastal State shall deposit with the Secretary-General of the United Nations charts and relevant information, including geodetic data, permanently describing the outer limits of its continental shelf. The Secretary-General shall give due publicity thereto.
10. The provisions of this article are without prejudice to the question of delimitation of the continental shelf between States with opposite or adjacent coasts.

Text of Article 76 of UNCLOS Copyright © 1983 United Nations.

1.2 The Irish National Seabed Survey (INSS)

Within the scope of the Irish National Seabed Survey and the successor project INFOMAR (Integrated Mapping For the sustainable development of the MArine Resource), Ireland has very nearly completed the Herculean task of mapping the entire Irish seabed, an area more than 9 times the size of the Republic of Ireland. While INFOMAR focused on the seabed shallower than 200 m bsl, the aim of the INSS was to map all areas deeper than 200 m bsl. The INSS was managed by the Geological Survey of Ireland (GSI) with the collaboration of the Irish Marine Institute (MI) while INFOMAR is a joint venture between the GSI and the MI.

Launched in 1999, the INSS represented one of the largest marine mapping programmes undertaken worldwide. During 6 years, from 1999 until 2005, the majority of the Irish continental marine area (413,867 km^2) was mapped by INSS and INFOMAR.

In numbers, the INSS represents a €32 million Irish investment that resulted in 5.5 terabyte total marine digital data sets of various types and over 300 paper-based charts. In order to collect these data, ships sailed a total of 250,815 km or 135,429 nautical miles (Fig. 1.4). The data collected during the INSS are all available online, free of charge and can be downloaded from http://www.infomar.ie/data/.

1.3 Seabed Features

The high resolution bathymetric data now available provides, for the first time, the necessary information required to map and investigate morphological seabed features tens and hundreds of metres wide. This means that not only can we find small-scale features but we can also map the larger features with much better accuracy.

The southern part of Ireland's seabed territory is characterised by large submarine canyons of which some are still active pathways for shelf to deep-sea exchange and submarine sediment avalanches called "turbidity flows". The canyons can be several hundred kilometres long, cutting hundreds of metres deep into the continental margin. They represent highways for the fast exchange of water masses, nutrients and sediments between the shelves and the deep sea (see Chapter 7) (Fig. 1.5).

In addition to submarine canyons, large channel systems occur on the deep Irish seabed. Like the canyons, the channels represent transport pathways for sediments and water masses. But rather than being deeply carved into the continental margin, channels meander down the more gently sloping stretches of the continental margin resembling large river systems in the terrestrial realm (see Chapter 7) (Fig. 1.6).

Other prominent features on the deep Irish seabed are seamounts defined as topographic highs (or hills) that elevate more than 1,000 m above the seafloor. They are mostly volcanic in origin and, due to special oceanographic conditions and hydrographic environments, seamounts represent areas of high species diversity and abundance. Because their summits are so isolated, seamounts often exhibit distinct and unique biological assemblages including a significant number of endemic species found nowhere else in the world. Seamounts are biological "hot spots" and also of economic importance for the fishing industry (see Chapter 8) (Fig. 1.7).

Escarpments, or submarine cliff faces, form as a result of strong and erosive bottom currents or are the scars left following submarine (land)slides. The deep oceans are not necessarily a quiet environment where sediments, once deposited, remain in place forever. Ocean circulation and strong currents occur in the deep ocean as much as in surface waters and even tides affect water mass boundaries several thousand metres below sea level generating so-called internal tides. Along the continental slopes especially, strong contour-parallel currents re-suspend, erode, scour, transport

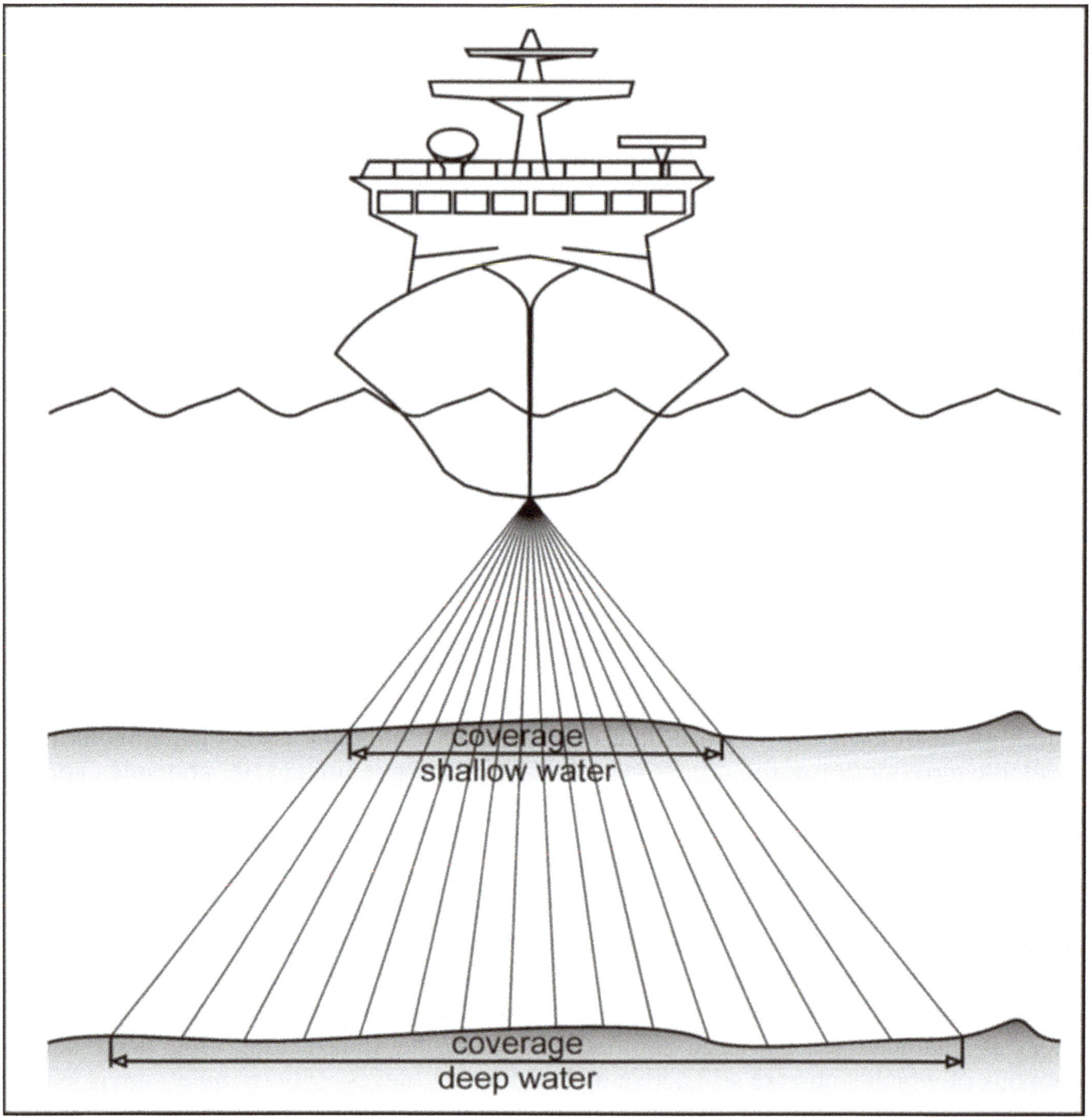

Fig. 1.3 Resolution verses coverage. The spacing of the mapping vessel's track lines depends on the water depth. The deeper the water the wider the swath width of the mapping coverage, however there is a pay-off, the wider the coverage the lower the resolution

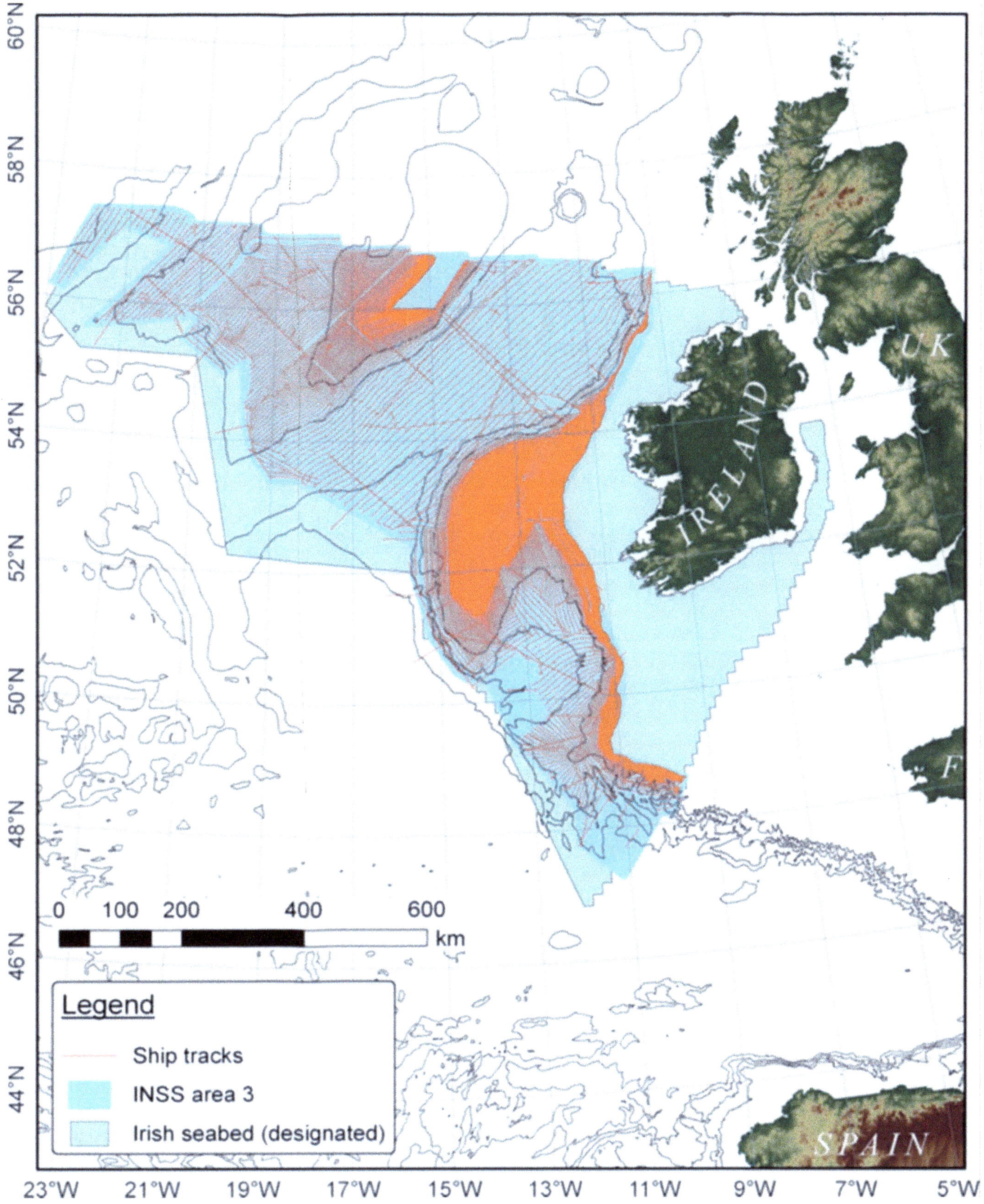

Fig. 1.4 INSS coverage. Ireland's underwater realm is huge relative to the landmass. This map shows the ship's tracks and highlights the tremendous effort necessary to receive a full coverage of the seabed with multi-beam echosounder. INSS area 3 is the seafloor below 200 m water depth. Bathymetric contours are based on GEBCO (General Bathymetric Chart of the Ocean) data

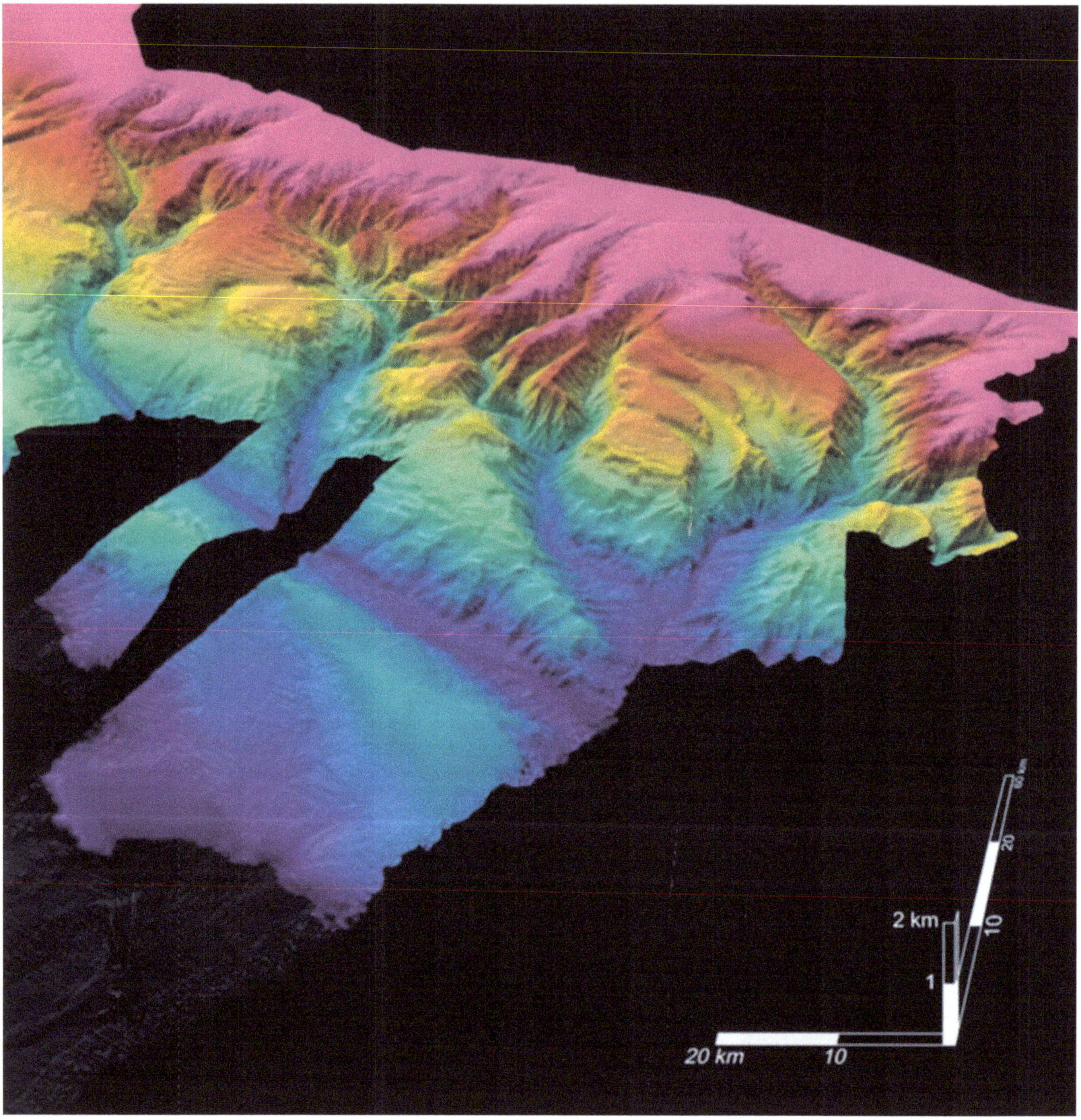

Fig. 1.5 Seabed features: Canyons. The Whittard Canyon system southwest of Ireland is one example of the many deep marine canyons that are carved into the Irish continental margin. This particular canyon system is approximately 300 km long, has formed during times of sea level low stand (glacials), connects the shelf with the abyssal plain and has been active since the mid Miocene (~15 million years ago). The Whittard Canyon system is a fast track for sediments from the Great Sole drainage area which connects the Irish Sea seabed to the Celtic Deep-Sea Fan or turbidite system. The image has a 6x vertical exaggeration (VE)

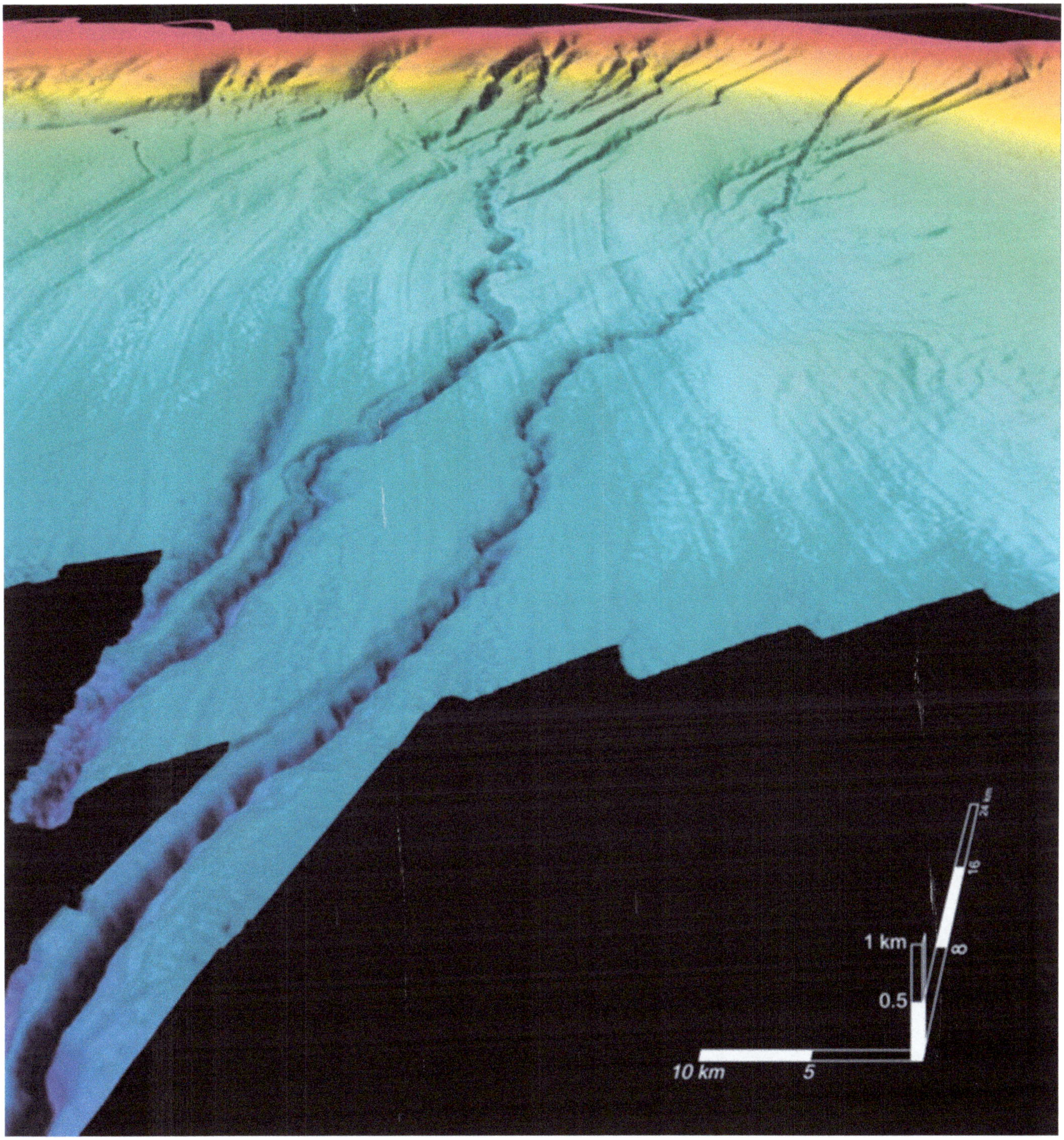

Fig. 1.6 **Seabed features: Channel systems.** The Gollum Channel system extends from the shelf break of the eastern Porcupine Seabight all the way through the mouth of the Seabight and onto the Porcupine Abyssal Plain. It is approximately 300 km long and nicely resembles the shape of a terrestrial meandering river system. The image has a 6x VE

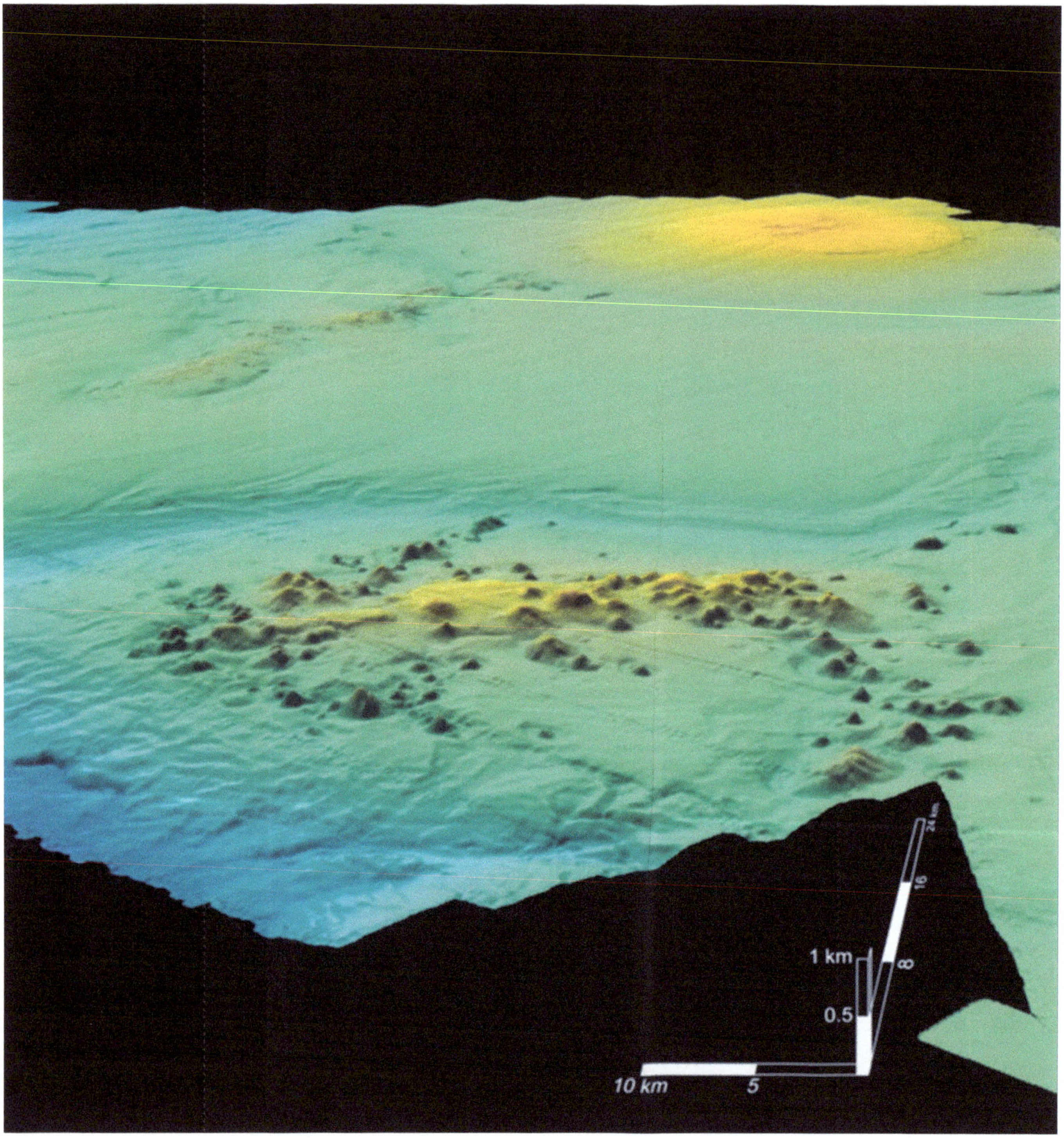

Fig. 1.7 Seabed features: Seamounts. An unnamed seamount in the southern Hatton-Rockall Basin. This seamount measures approximately 70 km across and 1,830 m from bottom to top (~2,550 to 720 m water depth). It is most likely of volcanic origin and has an irregular shaped summit. In the background, the southernmost tip of the Hatton Bank can be seen. The image has a 6x VE

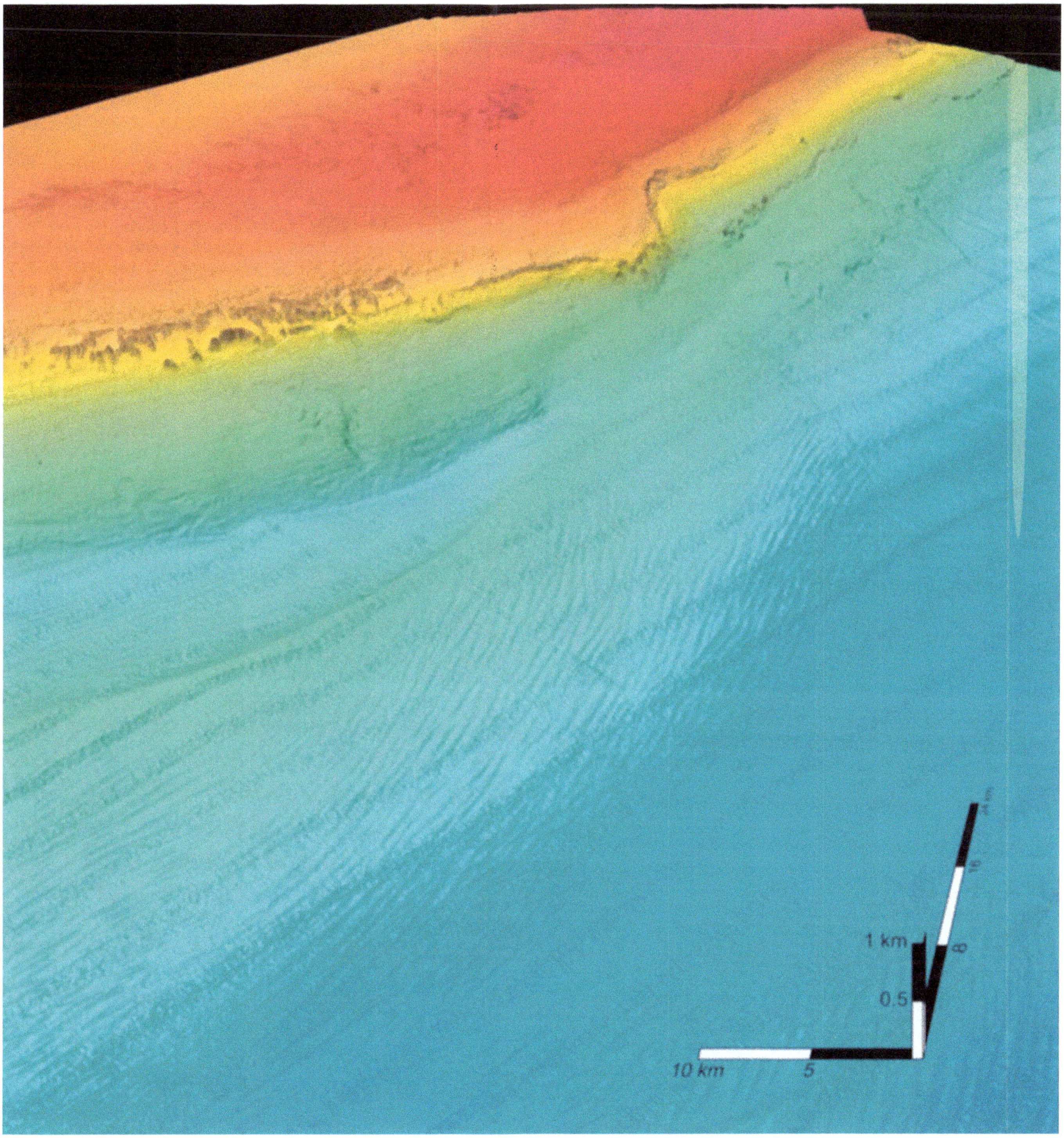

Fig. 1.8 **Seabed features: Escarpments.** Escarpments along the eastern Rockall Bank (*upper picture*) are steps in the seafloor caused by erosive bottom currents and sediment slumps and slides. Sediment waves (dune-like structures) formed by migrating sediment fill the foreground. At the left edge of the image, the start of the Logachev mound province can be seen (Fig. 1.9). The image has a 6x VE

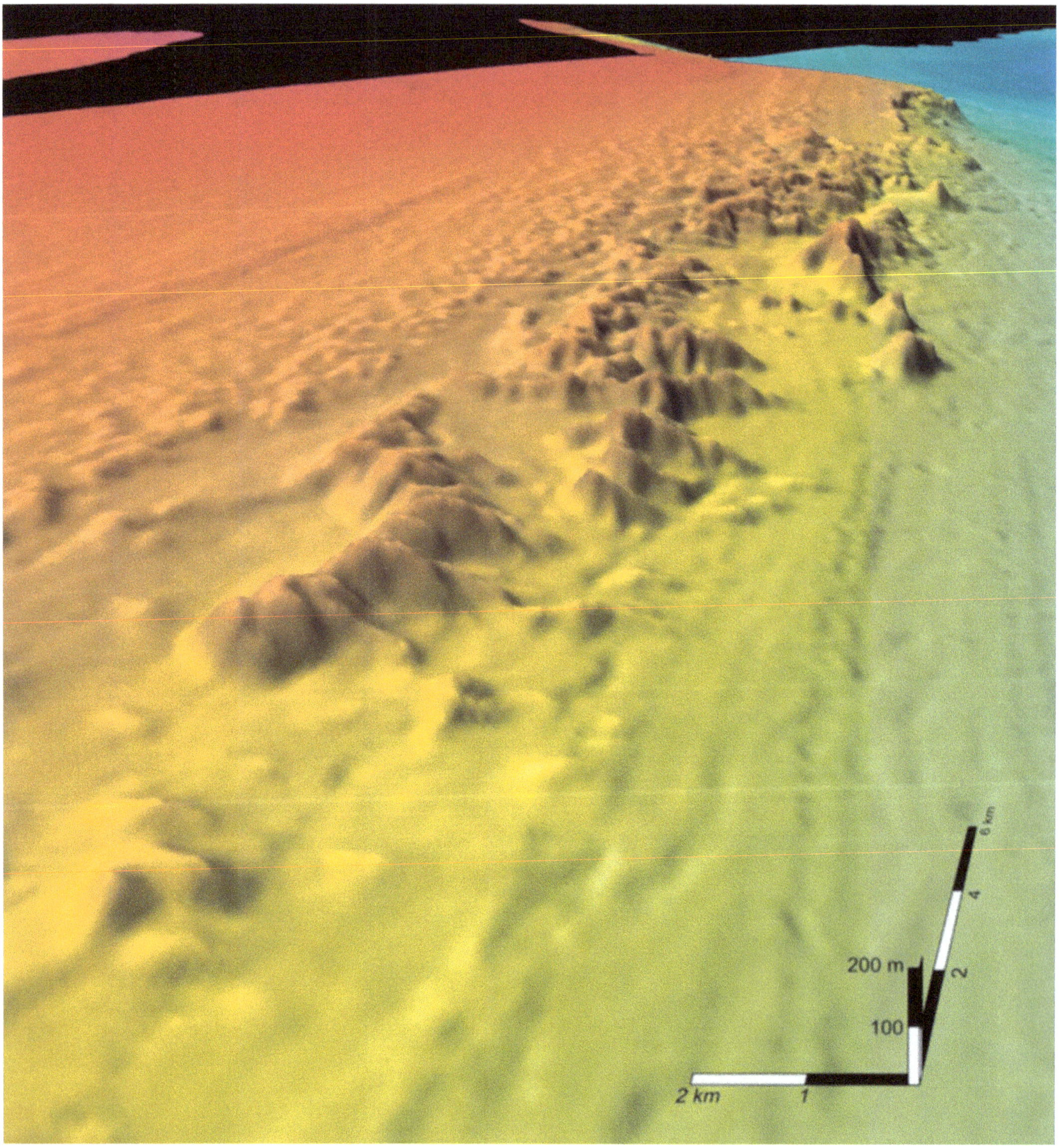

Fig. 1.9 Seabed features: Coral carbonate mounds. The Logachev mound province on the eastern Rockall Bank comprises approximately 70 coral carbonate mounds including some of the largest mounds known worldwide. The southernmost (*lower left*) large mound is ~5.5 km long and rises 390 m above the surrounding seafloor (950–560 m water depth). The image has a 6x VE

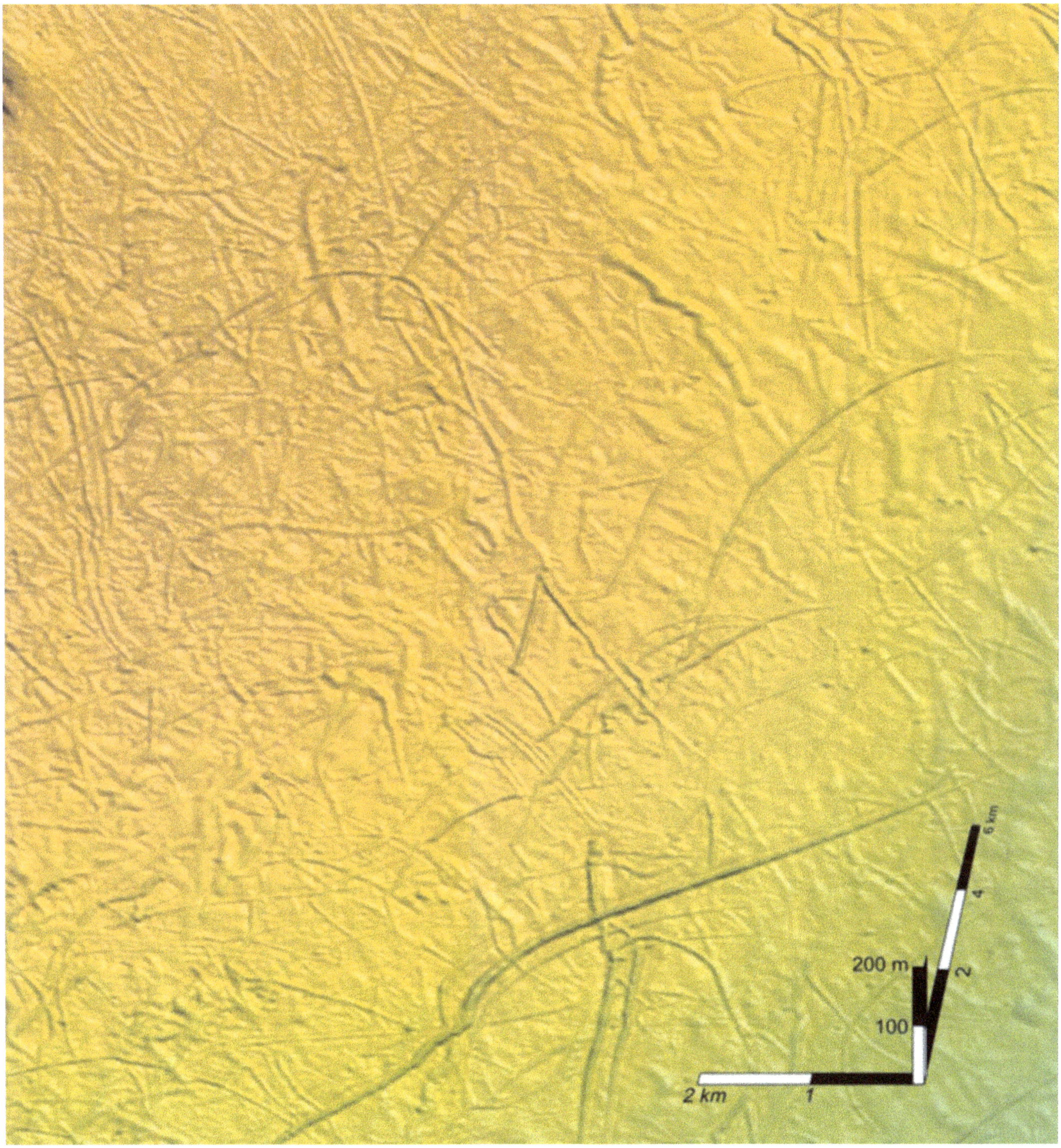

Fig. 1.10 Seabed features: Iceberg ploughmarks. Iceberg ploughmarks on the southern Porcupine Bank. They are caused by icebergs grounding on the upper continental margin and are common features on the entire Irish seabed shallower than approximately 500 m water depth. They were created during past glacials when icebergs travelled further south than today. In contemporary, warm, interglacial climatic conditions, icebergs no longer travel as far south as Ireland. The image has a 6x VE and a low angle of illumination

and eventually re-deposit sediments forming complex sedimentary patterns (see Chapter 9) (Fig. 1.8).

A close examination of the seabed also reveals many mound features that are much smaller than seamounts, down to tens of metres across, and can often, but not always, occur in swarms. These mounds may be of volcanic origin (mini seamounts), simple rock outcrops or biogenic structures such as cold-water coral reefs. In any case, these features usually offer a hard substrate that can be colonised by various organisms, for example cold-water corals, sea fans (gorgonians) and sponges. Such organisms provide further microhabitats that attract other species, and mounds are, therefore, often sites of relatively high biodiversity and species abundance (see Chapter 10).

Amongst these mound features are the cold-water coral carbonate mounds. It may come as a surprise that the Irish deep seabed hosts a plethora of coral reefs as we usually associate coral reefs with warm, tropical waters. In fact, the majority of coral species (65%) actually live in water depths deeper than 50 m. In Irish waters, cold-water corals have been reported from as far down as 2,000 m bsl. Like tropical corals, some species of these cold-water corals have the capacity to form reefs. In contrast to tropical reefs, however, Irish reefs are not found close to the shore but on the continental margin between 500 and 1,500 m bsl. They are, nevertheless, amazing ecosystems that in terms of species abundance and diversity are comparable with their tropical counterparts. The Irish deep-water seabed contains the largest and best studied examples of cold-water coral carbonate mounds in the world (see Chapter 10) (Fig. 1.9).

Another intriguing feature of the Irish seabed is a relic of past glaciation. Under contemporary, warm climatic conditions, icebergs do not travel as far south as Ireland. This was, however, quite different during past glacial periods when climatic conditions were much colder than today with icebergs travelled much further south, even further than Ireland. These icebergs, with 7/8th hidden below the water line, grounded on the continental margin scarring the seabed and leaving very characteristic so-called iceberg ploughmarks which are typically long furrows in the seabed zigzagging obliquely upslope on the upper continental margins. They can be found in all areas of the Irish seabed shallower than approximately 500 m bsl (see Chapter 11) (Fig. 1.10).

INSS Data Processing

2

B. Dorschel, A.J. Wheeler, X. Monteys, and K. Verbruggen

This seabed atlas is based on 55 gigabytes (more than 16.5 billion soundings) of processed bathymetric data that were collected during 18 cruises between 2000 and 2002 as part of the INSS. Data acquisition was carried out using a deep-water mutibeam echosounder system (Kongsberg-Simrad EM120) and two mid-water multibeam echosounder systems (Kongsberg-Simrad EM1002) onboard of the research vessels RV Bligh and RV Siren. After initial on board hydrographic processing with a Kongsberg-Neptune™ system the data are now available as web-based downloads from http://www.infomar.ie/ as text-files containing latitude, longitude and water depth (XYZ) coordinates, as digital elevation models (DEM) and raster data sets. For all geographic data presented in this atlas, the spatial reference datum is WGS1984 (Fig. 2.1).

Based on the XYZ coordinates, DEMs with 25 m resolution in the north-south direction were generated for this atlas using the IVS3D software package Fledermaus®.

B. Dorschel (✉) and A.J. Wheeler
Department of Geology & Environmental Research Institute, University College Cork, Cork, Ireland
e-mails: b.dorschel@ucc.ie; a.wheeler@ucc.ie

X. Monteys and K. Verbruggen
Geological Survey of Ireland, Beggars Bush, Haddington Road, Dublin 4, Ireland
e-mails: Xavier.Monteys@gsi.ie; Koen.Verbruggen@gsi.ie

Due to the large amount of data, it was necessary to subdivide the data set into 2° × 2° data blocks, each block measuring 222.2 km in the north-south direction and between 154.4 and 117.8 km in the east-west direction (due to the curvatures of the Earth). The data blocks are labelled 1 to 8 from west to east and A to F from north to south (Fig. 2.2).

The density of the multibeam data points is sufficient to model full coverage DEMs of 25 m resolution down to a water depth of approximately 2,000 m bsl (depending on morphology and slope angle). In water depths below 2,000 m bsl, the individual soundings are wider spaced hence the information density is lower. When DEMs are generated from areas with lower information density, outlying soundings increasingly influence the interpolated DEMs. The consequence is an increase in artefacts (data errors and outliers) for the deepest areas mapped. But even in the deepest part of the Irish seabed, artefacts are never too pronounced to overprint 100 m sized morphological features.

For further analyses, the DEMs generated in Fledermaus® were exported as text files and converted into ArcGIS® raster data sets. These ArcGIS® raster data sets were the basis for subsequent geo-statistical analyses.

B. Dorschel et al., *Atlas of the Deep-Water Seabed: Ireland*,
DOI 10.1007/978-90-481-9376-9_2, © Springer Science+Business Media B.V. 2010

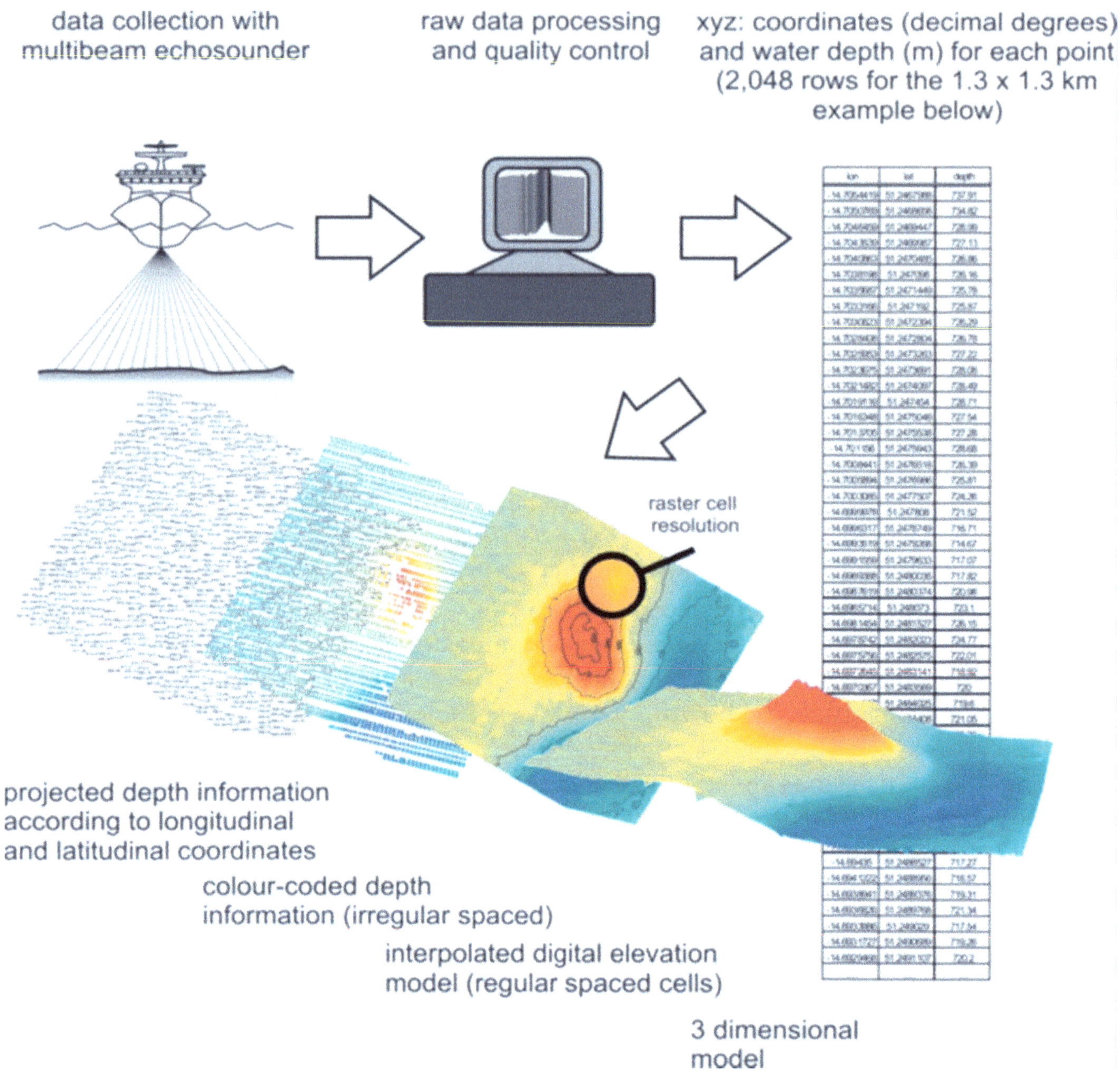

Fig. 2.1 **From sound to maps.** Electromagnetic waves such as light or radio waves do not travel far in water and therefore cannot be used to map the deep-water seabed. Sound waves, however, travel well in water and are used in various applications for mapping the seabed. Bathymetric data used for this atlas were recorded with a multibeam echosounder. This image shows the data processing progression (from *left* to *right*): collection at sea of individual soundings which are then cleaned, corrected and coded to produce an interpolated grid of bathymetric data. The gridded data are plotted producing a raster image which can then be manipulated to produce three-dimensional models and other outputs

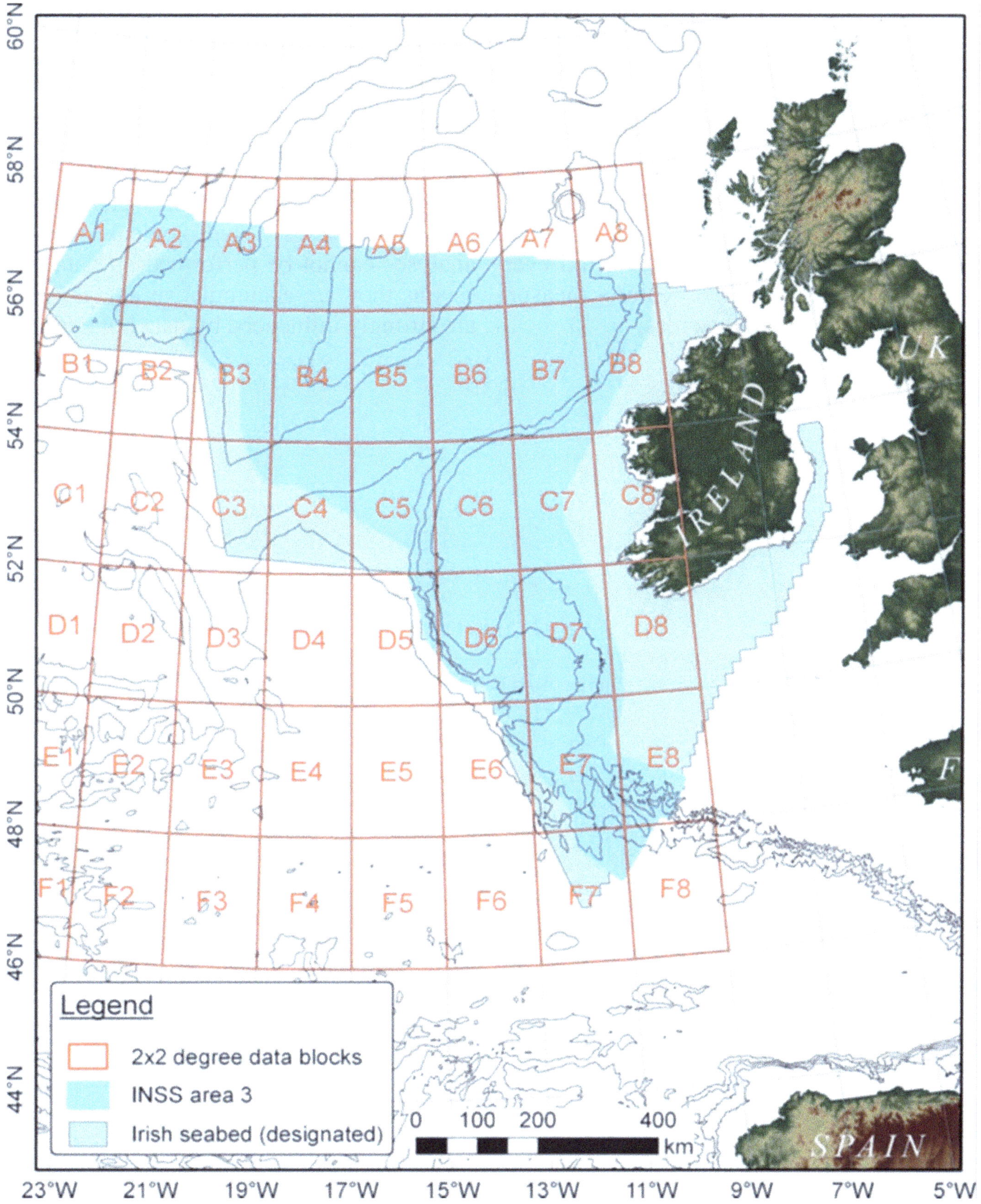

Fig. 2.2 Data blocks. The resolution of the bathymetry data and the size of the Irish seabed made it impossible to process all data sets in one go. The entire data set was therefore subdivided into 2° × 2° data blocks. These data blocks are easier to handle whilst still providing the full resolution of the initial data. Bathymetric contours are based on GEBCO data

These DEMs allowed, for the first time, a high resolution study of seabed features on the entire Irish deep-water seabed. The level of detail highlighted seabed features such as erosional scarps and cliffs and entire submarine canyons that had not been seen before and did not appear on charts. The internal structures of canyons are of almost artistic beauty and the level of detail even allows for the detection of features measuring only tens of metres across.

In addition to the high resolution raster data sets, an integrated raster was generated for the entire extent of multibeam coverage. The cell size of this raster is approximately 170 m in the north-south direction. This raster formed the base for hydrological network analyses in order to identify canyon and channel systems. Hydrological network analyses cannot be performed on data subsets as they are accumulative analyses that are strongly influenced by data boundaries.

3 The Irish Seabed

B. Dorschel, A.J. Wheeler, X. Monteys, and K. Verbruggen

The Irish seabed stretches from as far north as 58° 00′N on the Hatton Bank to as far south as 46° 45′S at the South-West Approaches and extends from 5° 16′W in the Irish Sea to the margin of the Iceland Basin at 24° 46′W. This atlas focuses on the part of the Irish seabed deeper than 200 m bsl covering an area of 536,112 km^2 west and south of Ireland.

The Irish seabed is defined as the extension of the continental crust that underlies Ireland and continues westward below the northeast Atlantic (see also chapter 1.1). The overall morphology of the Irish seabed is structured into a series of north-east/southwest orientated banks and basins (Fig. 3.1).

If we could dive across the Irish seabed on a virtual journey starting in southwest Ireland and heading towards Iceland, we would cross all major morphological seabed features (Fig. 3.2). The first part of the journey would cross the shallow Celtic Sea shelf. Beyond the shelf, one would encounter the Porcupine Seabight, a tongue-shaped embayment into the Celtic Margin, bordered in the west by the Porcupine Bank. The Porcupine Seabight developed as a failed rift during the opening phase of the Atlantic. During the descent down the eastern flank of the Porcupine Seabight, we would see several small mound structures. These structures are true Irish coral reefs, so-called cold-water coral carbonate mounds. The Gollum Channel would be to our left, a large channel system that meanders down the eastern flank of the Porcupine Seabight, continues across the bottom and through the mouth of the Seabight and onto the Porcupine Abyssal Plain.

After a gentle ascent, the western flank of the Porcupine Seabight or the eastern flank of the Porcupine Bank respectively, the Rockall Trough and the Rockall Bank, with the Hatton Bank in the distance would lie before us. A steep climb down the western flank of the Porcupine Bank would bring us to the bottom of the Rockall Trough. During the first part of our descent, more cold-water coral carbonate mounds would be seen as well as several vertical cliffs, so-called escarpments. Deep canyons at irregular spaced intervals can be found to the left and right.

An alternative route could be taken out of the mouth of the Porcupine Seabight to the southwest and onto the Porcupine

B. Dorschel (✉) and A.J. Wheeler
Department of Geology & Environmental Research Institute, University College Cork, Cork, Ireland
e-mails: b.dorschel@ucc.ie; a.wheeler@ucc.ie

X. Monteys and K. Verbruggen
Geological Survey of Ireland, Beggars Bush, Haddington Road, Dublin 4, Ireland
e-mails: Xavier.Monteys@gsi.ie; Koen.Verbruggen@gsi.ie

B. Dorschel et al., *Atlas of the Deep-Water Seabed: Ireland*,
DOI 10.1007/978-90-481-9376-9_3, © Springer Science+Business Media B.V. 2010

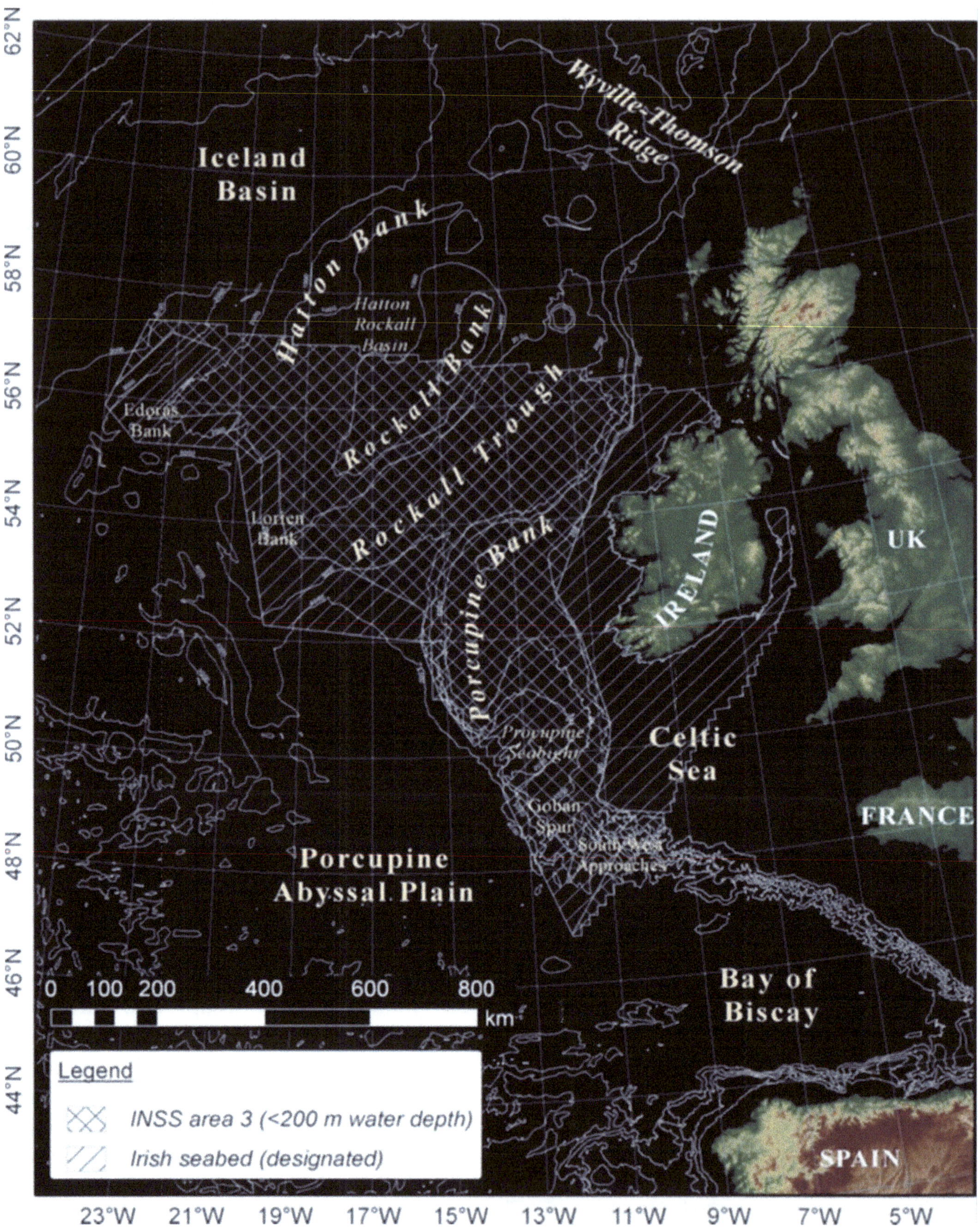

Fig. 3.1 Overview of the large morphological seabed features. This overview figure highlights the various banks, troughs, and seamounts in and around the Irish continental shelf. It shows that the ocean floor is not flat and even like a pan but is warped into valleys, troughs, hills, cliffs and mountain ranges akin to sub-aerial landscapes. And just as in the terrestrial realm, the different landscapes / types of seabed and altitudes / bathymetries create different habitats with different biological assemblages. Bathymetric contours are based on GEBCO data

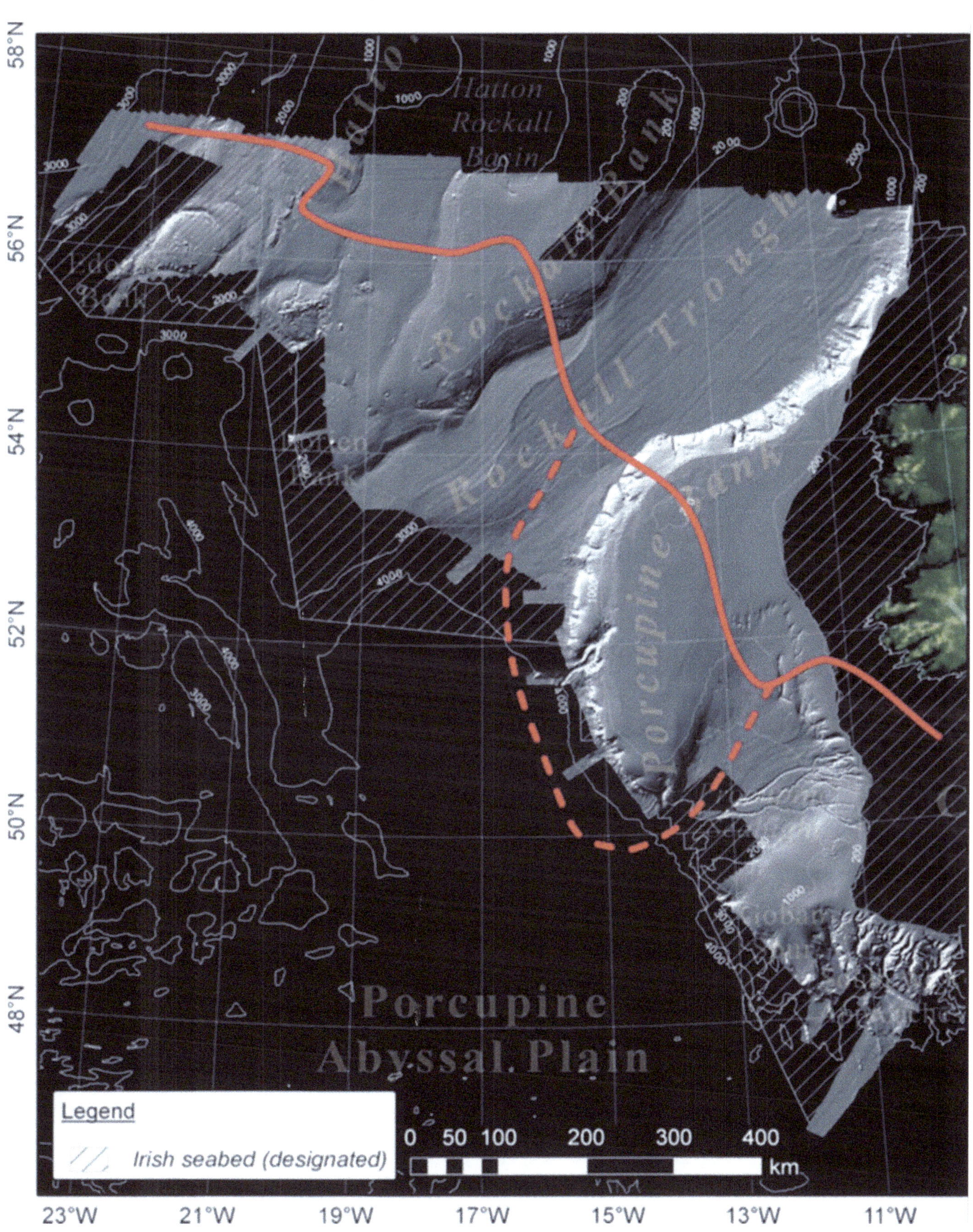

Fig. 3.2 Overview of the deep-water Irish seabed. Hillshade image of the Irish seabed deeper than 200 m water depth. This image nicely illustrates how the seabed would look like without water and with the sun shining from the north-west. The route of our "virtual dive across the seabed" described in the chapter text is shown as a red line. The dashed line indicates the detour over the Porcupine Abyssal Plain

Abyssal Plain at ~5,000 m water depth, and then northwards into the Rockall Trough.

The Rockall Trough is an area of thinned but continuous continental crust between the Porcupine and the Rockall Bank. Like the Porcupine Seabight, the Rockall Trough also developed as a series of failed rifts during the opening phases of the Atlantic. The Rockall Trough deepens from the north-east to south-west to ~4,000 m water depth and also opens onto the Porcupine Abyssal Plain. Morphologically, the Rockall Trough is characterised by a relatively even topography. The only slight obstacle to be crossed on the journey would be the Feni Ridge, a body of drift sediments in the western Rockall Trough.

From the bottom of the Rockall Trough there is a climb of more than 2,000 m onto the Rockall Bank. The Rockall Bank is, like the Porcupine Bank, another continental block which split-off during the opening phases of the north Atlantic.

Climbing up the Rockall Bank, more cold-water coral carbonate mounds and escarpments would be seen. These cold-water coral carbonate mounds are the largest known worldwide and many are covered by dense thickets of living coral colonies. At the escarpments, the seabed often rises almost vertically for up to 300 m. Canyons and channels, however, are mainly absent. With two small exceptions, there are no canyons on the eastern Rockall Bank.

After the climb up the Rockall Bank, the tallest slopes are left behind. Towards the west, the Rockall Bank dips gently into the shallower Hatton-Rockall Basin. On this flank, hundreds of small (approximately 50 m high) cold-water coral carbonate mounds are found scattered over the seabed.

Crossing the Hatton-Rockall Basin, a few channel systems are found before encountering the southern tip of the Hatton Bank. The Hatton Bank also represents a split-off block of continental crust and the Hatton-Rockall Basin, like the Rockall Trough, is underlain by thinned continental crust. To the south of the Hatton Bank are impressive seamounts with superimposed smaller mound features.

Descending the Hatton Bank towards the west and into the Iceland Basin, the Maury Channel is encountered and, leaving the Irish seabed, we finally move from continental crust onto oceanic crust.

4 Below the Irish Seabed

B. Dorschel, A.J. Wheeler, X. Monteys, and K. Verbruggen

Three hundred million years ago the world looked quite different with Ireland at the centre of a continent known as Pangea. It was at about this time that Pangea started to split apart creating a series of rift valleys. Eventually, this rifting split all the way through the continental crust and a new ocean basin was formed with two new smaller continents on either side: the North Atlantic, Europe and North America respectively. The gross morphological structures of the Irish deep-water seabed developed as a result of a series of rift valleys that formed during the early development of the north Atlantic starting in the Permo-Triassic geological period (around 250 million years ago).

The Porcupine Seabight, Rockall Trough and Hatton Basin are the rift valleys where the Atlantic first tried to open. The eventual opening of the Atlantic occurred further to the west with spreading still going on today centred on Iceland and the Mid-Atlantic ridge. Between the rifts are the remnants of the Pangean continental blocks that now form the Porcupine, Rockall and Hatton Banks. These nearly reach to the sea surface (in fact a tiny part of the northern Rockall Bank, the Rockall Islet, pokes just above the sea). Although the continental crust was thinned in the troughs during rifting, no new oceanic crust was formed and this whole area is therefore still part of the European continental crustal block even though it is submerged underwater (Fig. 4.1).

The rocks that make up the Porcupine, Rockall and Hatton Banks are very old consisting in part of Pre-Cambrian and Lower Palaeozoic rocks (900–400 million years old) and Caledonian granites (ca. 400 million years old), very similar to some of the rocks we find on land in Ireland. These continental blocks can be up to 28 km thick and are overlain by younger sediments. During rifting, the crust in the troughs thinned to as little as 6 km. These troughs have then become progressively filled with several kilometres of sediments (Fig. 4.2). Importantly, the sediment infill at the edges of the troughs contains hydrocarbons, such as the natural gas reservoir known as the Corrib Field.

The structure of the Irish seabed is dominated by rift structures with fault zones

B. Dorschel (✉) and A.J. Wheeler
Department of Geology & Environmental Research Institute, University College Cork, Cork, Ireland
e-mails: b.dorschel@ucc.ie; a.wheeler@ucc.ie

X. Monteys and K. Verbruggen
Geological Survey of Ireland, Beggars Bush, Haddington Road, Dublin 4, Ireland
e-mails: Xavier.Monteys@gsi.ie; Koen.Verbruggen@gsi.ie

B. Dorschel et al., *Atlas of the Deep-Water Seabed: Ireland*,
DOI 10.1007/978-90-481-9376-9_4, © Springer Science+Business Media B.V. 2010

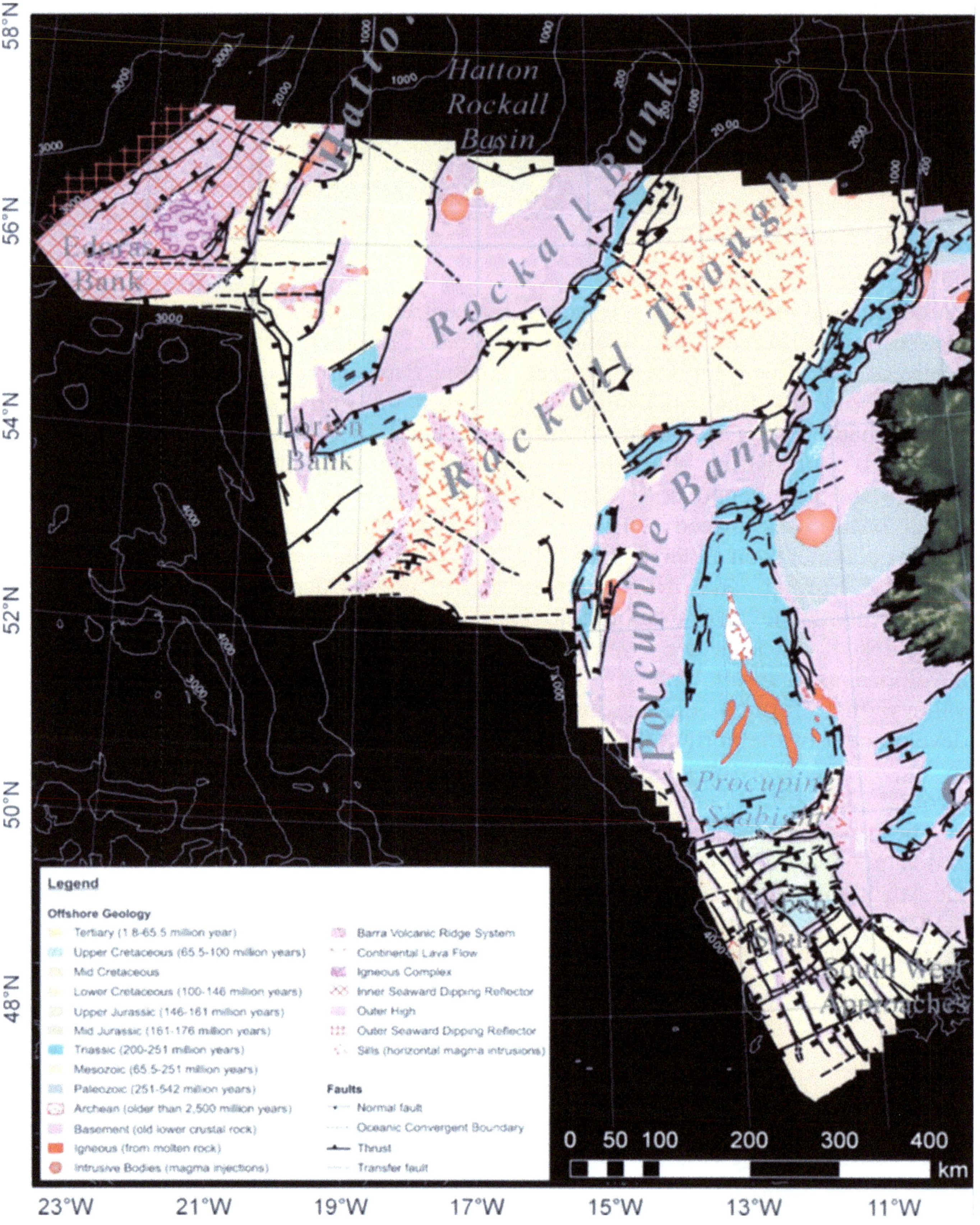

Fig. 4.1 The geology of the Irish seabed. The geological map of the Irish seabed highlights the banks as crystalline basement (old continental crust) whilst the trough and basins were successively filled with sediments. The orientation of faults is perpendicular to the direction of expansion of the crust during the opening phase of the Atlantic. (Petroleum Affairs Division/DCENR website (offshore geology compiled by PAD technical staff from various sources))

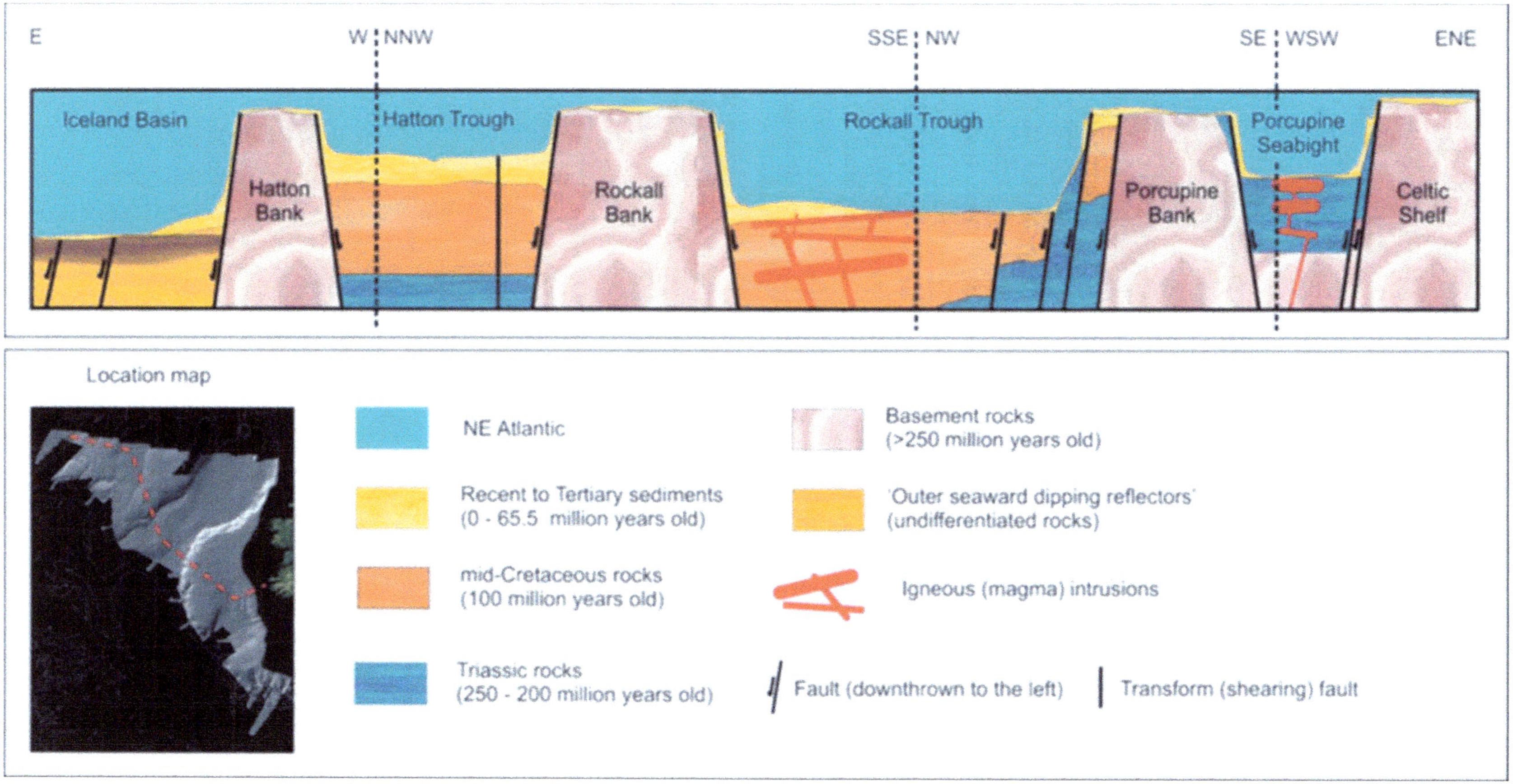

Fig. 4.2 **Idealised cross-section across the seabed.** The main troughs (Hatton and Rockall Trough and Porcupine Seabight) are rifts bounded by faults, downthrown as the area was pulled apart during the early opening of the north Atlantic basin. The banks (Hatton, Rockall and Porcupine) and the Celtic Shelf are remnants of the former continent of Pangea. Injections of magma from past volcanic activity form sills, dykes and seamounts in the troughs

delineating the boundaries between the troughs and banks. The faults were where earthquake activity was focused as the Atlantic tried to open. As the spreading of the Atlantic is now focussed on Iceland, the Irish seabed is now, thankfully, tectonically stable. The numerous faults are so-called 'normal' faults which represent extensional fractures. There are also a few large transform faults where a shearing motion has fractured the crust causing the rocks to rub past each other. These were formed during this complex, early Atlantic opening phase as well (Fig. 4.2).

During the active rifting phase when the crust was pulled apart, the thinning of the crust was accompanied by volcanic activity. As the crust thins, there is less pressure at (or weight on) the base of the crust which helps it to melt producing magma. This rises up through the thinning crust and may even have erupted onto the seafloor. Similar activity is presently happening in Iceland and on its surrounding seabed. The numerous seamounts on the Irish seabed are evidence of this process as are the numerous buried igneous rocks (formed from molten rock) such as basalt sills, dykes and young granites.

Bibliography

Bailey, W., Shannon, P. M., Walsh, J. J., & Unnithan, V. (2003). The spatial distribution of faults and deep sea carbonate mounds in the Porcupine Basin, offshore Ireland. *Marine and Petroleum Geology, 20*, 509–522.

Croker, P. F., & Shannon, P. M. (Eds.). (1995). *The petroleum geology of Ireland's offshore basins, 93*. London, Special Publications, London: Geological Society. 493 pp.

Shannon, P. M. (1991). The development of Irish offshore sedimentary basins. *Journal of the Geological Society London, 148*, 181–189.

Shannon, P. M., Haughton, P. D. W., & Corcoran, D. V. (2001). *The Petroleum Exploration of Ireland's Offshore Basin*. London: The Geological Society.

Shannon, P. M., McDonnell, A., & Bailey, W. R. (2007). The evolution of the Porcupine and Rockall basins, offshore Ireland: the geological template for carbonate mound development. *International Journal of Earth Sciences, 96*(1), 21–35.

5 On the Irish Seabed

B. Dorschel, A.J. Wheeler, X. Monteys, and K. Verbruggen

In addition to the large scale morphological features, the Irish seabed also hosts numerous smaller features such as canyons, seamounts, escarpments and mound features including coral carbonate mounds and iceberg ploughmarks. Many of these features are so small that they do not show up on standard overview maps. As a result, up until very recently, only patchy information for some areas of scientific interest has been available. The INSS data now allows for a comprehensive mapping of these features including small mounds that may only rise several metres above the seabed (Fig. 5.1).

B. Dorschel (✉) and A.J. Wheeler
Department of Geology & Environmental Research Institute, University College Cork, Cork, Ireland
e-mails: b.dorschel@ucc.ie; a.wheeler@ucc.ie

X. Monteys and K. Verbruggen
Geological Survey of Ireland, Beggars Bush, Haddington Road, Dublin 4, Ireland
e-mails: Xavier.Monteys@gsi.ie; Koen.Verbruggen@gsi.ie

B. Dorschel et al., *Atlas of the Deep-Water Seabed: Ireland*,
DOI 10.1007/978-90-481-9376-9_5, © Springer Science+Business Media B.V. 2010

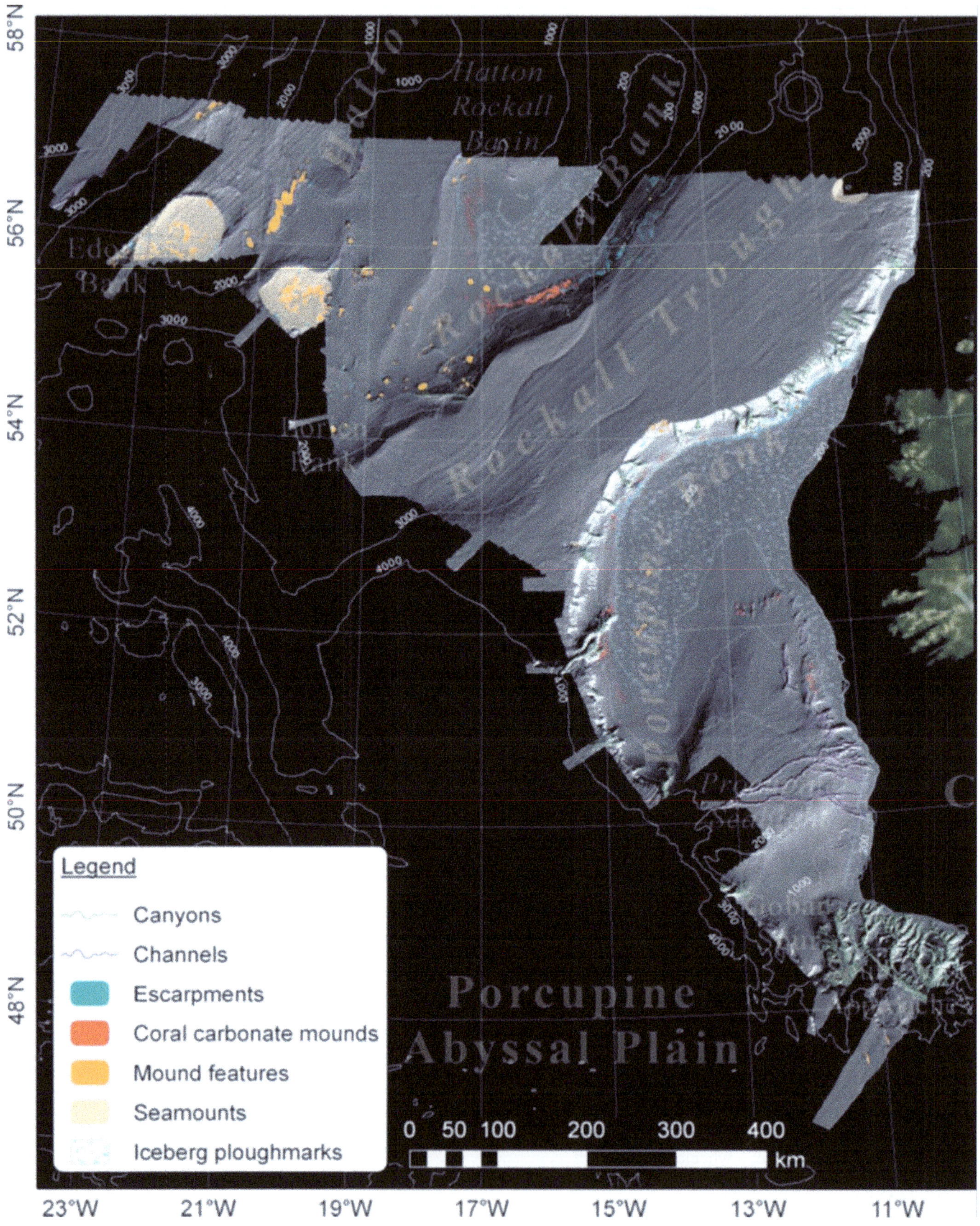

Fig. 5.1 **Seabed features: Distribution.** This map illustrates the multibeam echosounder coverage (grey hillshade bathymetry relief) and provides the geographical context for the morphological features highlighted in this atlas. The numerous canyon and channel systems along the eastern Irish continental margin can be seen as well as the two large seamounts in the west. Coral carbonate mounds and escarpments are also evident, aligned along the upper continental margin with extensive areas of iceberg ploughmarks on the upper banks

Above the Irish Seabed

6

B. Dorschel, A.J. Wheeler, X. Monteys, and K. Verbruggen

The ocean above the Irish seabed is by no means a homogeneous body of water. Like most areas of the world's oceans, it is layered with distinct water-masses occurring at distinct water depth intervals. As a result of this stratification, the water in the ocean can flow in different directions at different depths. Individual water-masses can be characterised by their physical properties with the temperature and salinity (concentration of salt in seawater) properties of water-masses widely used for oceanographic research. Temperature and salinity (Fig. 6.1) are the so-called conservative properties of seawater as they remain relatively stable over time, only changing with the mixing of different water masses or with contact to the atmosphere. In addition to the conservative properties, numerous non-conservative properties are analysed like e.g. oxygen and carbon dioxide concentrations in the oceans.

Once decoupled from the atmosphere, water-masses keep their conservative properties during their journeys through the world's oceans. Water that was formed and downwelled in the Arctic, for example, can be traced, based on its properties, throughout the entire Atlantic all the way to Antarctica where it mixes with other water-masses to finally become distributed into the Indian and Pacific Oceans. On their journeys, water-masses occupy the water depths defined by their densities. In a stable, stratified ocean, the water-mass with the highest density forms the bottom water. This is overlain by the water-mass with the next highest density and so on. The water-mass density is controlled by temperature and salinity and therefore also conservative. Throughout an individual water-mass, density tends to be homogeneous. Between water masses, the density can change rapidly resulting in a steep vertical density gradients, so-called pychnoclines. It is along these pychnoclines that internal tides can occur similar to the tides felt at the sea surface.

In Irish waters, the most prominent water-masses are the Labrador Seawater (LSW), the Eastern North Atlantic Water (ENAW), the Sub-Arctic Intermediate Water (SAIW), the Mediterranean Outflow Water (MOW) and the Shelf Edge Current (SEC).

B. Dorschel (✉) and A.J. Wheeler
Department of Geology & Environmental Research Institute, University College Cork, Cork, Ireland
e-mails: b.dorschel@ucc.ie; a.wheeler@ucc.ie

X. Monteys and K. Verbruggen
Geological Survey of Ireland, Beggars Bush, Haddington Road, Dublin 4, Ireland
e-mails: Xavier.Monteys@gsi.ie; Koen.Verbruggen@gsi.ie

B. Dorschel et al., *Atlas of the Deep-Water Seabed: Ireland*,
DOI 10.1007/978-90-481-9376-9_6, © Springer Science+Business Media B.V. 2010

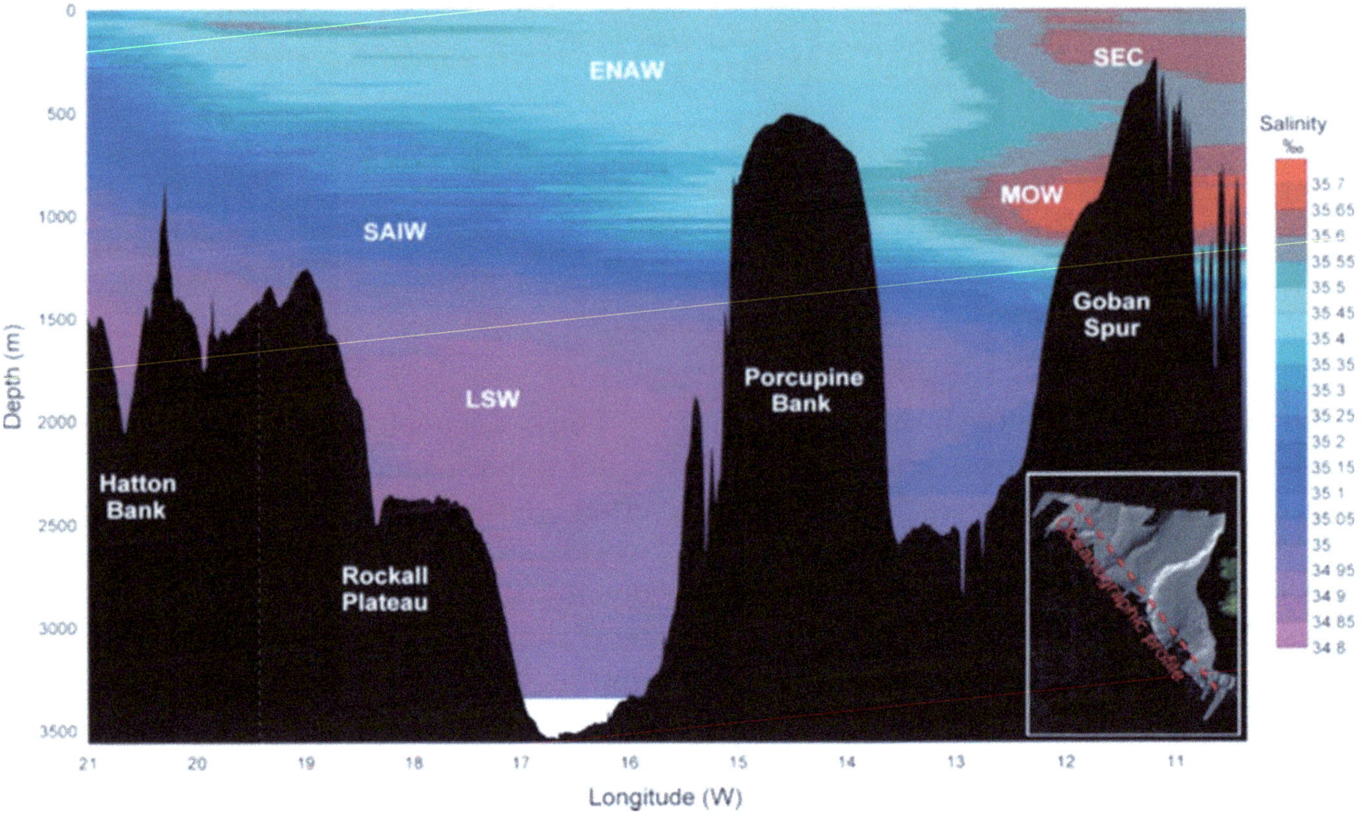

Fig. 6.1 The oceanography of Irish waters. The oceanographic overview illustrates the complexity of the ocean structure. During the INSS mapping program several hundred of temperature and salinity profiles were acquired. The chosen transect runs for a 1,000 km from the Hatton Bank to south of the Goban Spur and displays a subset 25 hydrographic stations. Based on this salinity plot several water masses can be identified. Labrador Sea Water (LSW) can be found in the deep trough and basins below approximately 800 m bsl. Above the LSW, Sub-Arctic Intermediate Water (SAIW) and Eastern North Atlantic Water (ENAW) can be found. Towards the eastern continental slope Mediterranean Outflow Water (MOW) occurs, thinning out from South to North at water depths between 600 and 1,100 m bsl. In the shallowest regions, the Shelf Edge Current (SEC) is sweeping northward along the upper part of the continental slope reaching up to the shelf

LSW occupies the deepest parts of the troughs and basins with its centre at approximately 1,700 m water depth. It forms, as the name implies, in the Labrador Sea and flows south-eastward across the Atlantic. Above the LSW, shallower than approximately 800 m bsl, is the SAIW in the west and the ENAW in the east. The ENAW forms in the Bay of Biscay and flows northward on top of the LSW. Further to the east, the MOW is found which flows out of the Mediterranean Sea through the Straits of Gibraltar, sinks to its density defined depth interval (in Irish waters, the centre of the MOW is at approximately 950 m bsl) and travels northward along the European continental slope. The MOW enters the Porcupine Seabight and can be traced as far north as the southern Rockall Trough. In the central and northern Rockall Trough, the MOW cannot be distinguished any more. The SEC represents a water-mass of mixed origins that flows northward along the upper continental slope up to the shelf edge.

Bibliography

Broecker, W. S. (1990). Salinity history of the nothern Atlantic during the last deglaciation. *Paleoceanography, 5*(4), 459–467.

Greatbatch, R. J. (2000). The North Atlantic oscillation. *Stochastic Environmental Research and Risk Assessment, 14*(4/5), 213–242.

Hargreaves, P. M. (1984). The distribution of Decapoda (Crustacea) in open ocean near bottom over an adjacent slope in the Northern North-east Atlantic Ocean during autumn 1979. *Journal of the Marine Biological Association of the United Kingdom, 64*, 829–857.

Harvey, J. (1982). θ-S relationships and water masses in the eastern North Atlantic Deep Sea Research Part A. *Oceanographic Research Papers, 29*(8), 1021–1033.

Prins, M. A., Bouwer, L. M., Beets, C. J., Troelstra, S. R., Weltje, G. J., Kruk, R. W., et al. (2002). Ocean circulation and iceberg discharge in the glacial North Atlantic: Inferences from unmixing of sediment size distributions. *Geology, 30*(6), 555–558.

van Aken, H. M. (2000). The hydrography of the mid-latitude Northeast Atlantic Ocean II: The intermediate water masses. *Deep-Sea Research I, 47*, 789–824.

van Aken, H. M., & Becker, G. (1996). Hydrography and through-flow in the north-eastern North Atlantic; the NANSEN project. *Progress in Oceanography, 38*, 297–346.

White, M., & Bowyer, P. (1997). The shelf-edge current north-west of Ireland. *Annales Geophysicae, 15*, 1076–1083.

Part II

Thematic Atlas

Canyons and Channels

7

B. Dorschel, A.J. Wheeler, X. Monteys, and K. Verbruggen

Submarine canyons are the most prominent morphological features on the Irish seabed. They are tens to hundreds of kilometres long, narrow valleys carved tens to several hundreds of metres deep into the continental margin. They often extend from the shelf break at 200 m water depth all the way down to the lower continental rise at approximately 2,500 m water depth funnelling large volumes of sediment and organic matter from shelf regions to the deep ocean (Fig. 7.1).

Canyons were most active during glacial times when thick ice caps covered northern Europe. These ice caps flowed downhill under their own weight and fed into glaciers. The movements of glaciers and ice caps are incredibly destructive. These most powerful agents of erosion literally grind down landscapes and gouge out valleys. When glaciers melt, they release huge volumes of sediment that may be dumped directly into the sea or transported via large swollen rivers. As large volumes of water are stored in ice caps rather than flowing back to the sea, sea level can drop to 120 m below present day sea level during glaciations. This means that all the sediments from the melting glaciers enter the sea near the continental margin and often flow straight into the canyon heads through which sediment-laden currents cascade. During interglacial periods (when sea levels are higher) canyons are relatively inactive and can become partly filled with hemipelagic sediments (sediments of terrigenous and biogenic components that are typical found on continental margins).

In many instances, contemporary canyons are directly linked to river systems on land but on the Irish seabed this is not the case. The Irish canyons in the south (Fig. 7.3) were linked to the large river systems that used to flow across the Celtic Sea seabed (when sea level was lower) fed by the melting British and European ice sheets. The substantial canyons on the Porcupine Bank are more of a paradox (Fig. 7.9 & 7.16). It is hard to envisage a connection to any major river system in the hinterland and there is no evidence that the European ice caps extended onto the Porcupine Bank during glacial periods. The heads of the Porcupine Bank canyons are also located in deep water usually below 950 m bsl with no evidence for a connection to the top of the bank. The origin and development of the Porcupine Bank canyons has not yet

B. Dorschel (✉) and A. Wheeler
Department of Geology & Environmental Research Institute, University College Cork, Cork, Ireland
e-mails: b.dorschel@ucc.ie; a.wheeler@ucc.ie

X. Monteys and K. Verbruggen
Geological Survey of Ireland, Beggars Bush, Haddington Road, Dublin 4, Ireland
e-mails: Xavier.Monteys@gsi.ie; Koen.Verbruggen@gsi.ie

B. Dorschel et al., *Atlas of the Deep-Water Seabed: Ireland*,
DOI 10.1007/978-90-481-9376-9_7, © Springer Science+Business Media B.V. 2010

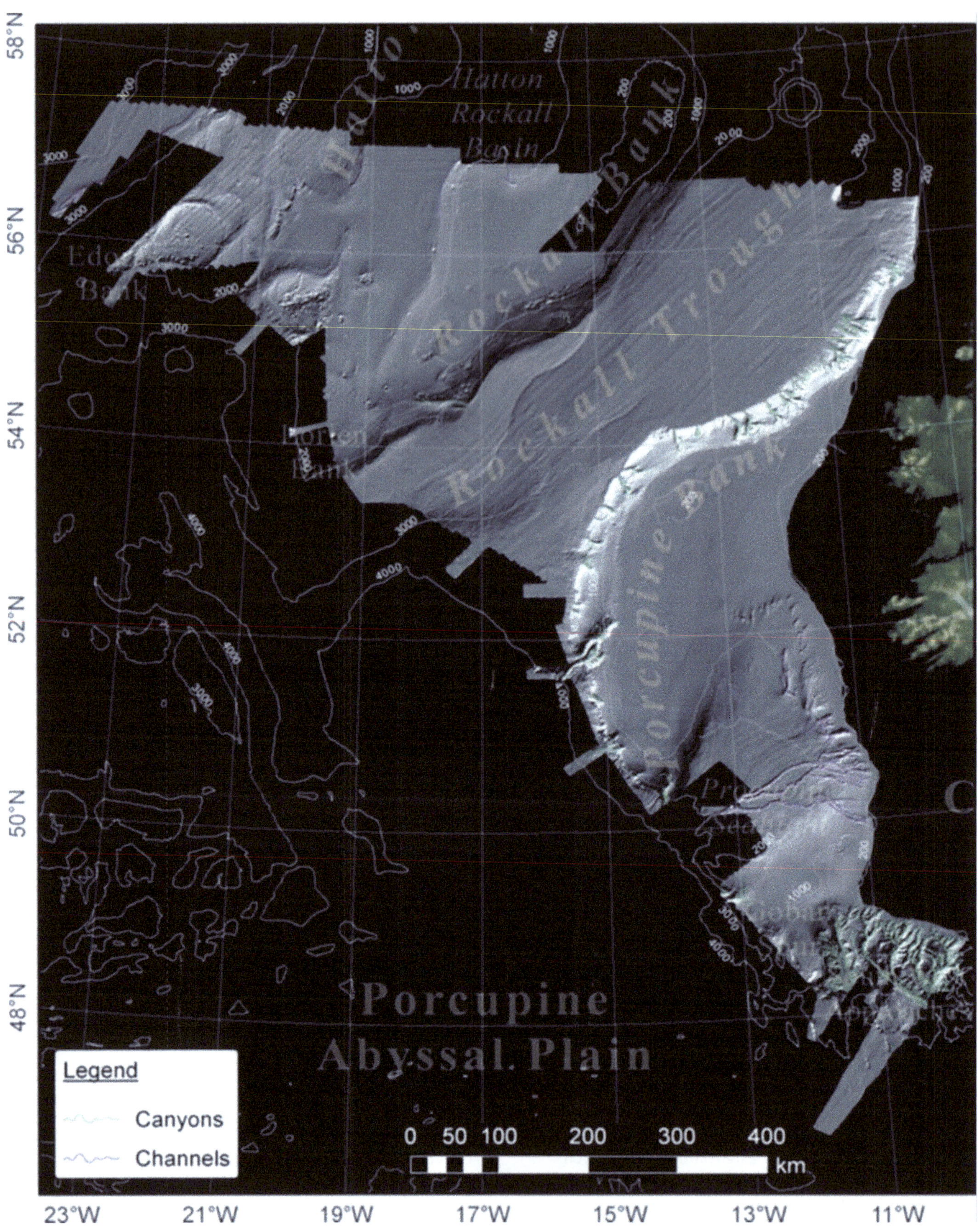

Fig. 7.1 **Canyon and Channels.** Canyons occur on both sides of the Rockall Trough and further south in the SW Approaches. Except for three small canyons at the east Rockall Bank, the Rockall Trough canyons are carved into the eastern continental margin. In the south, the canyons start more or less at the shelf break while the canyons along the Porcupine Bank and northernmost Irish continental margin barely cross the 1,000 m water depth contour. Channels can be found in the Porcupine Seabight and far to the west, south of Hatton Bank. Like the canyons of the SW Approaches, the channel system in the Porcupine Seabight reaches up to the shelf break. The western channels are deep seated in water depths below 1,000 m

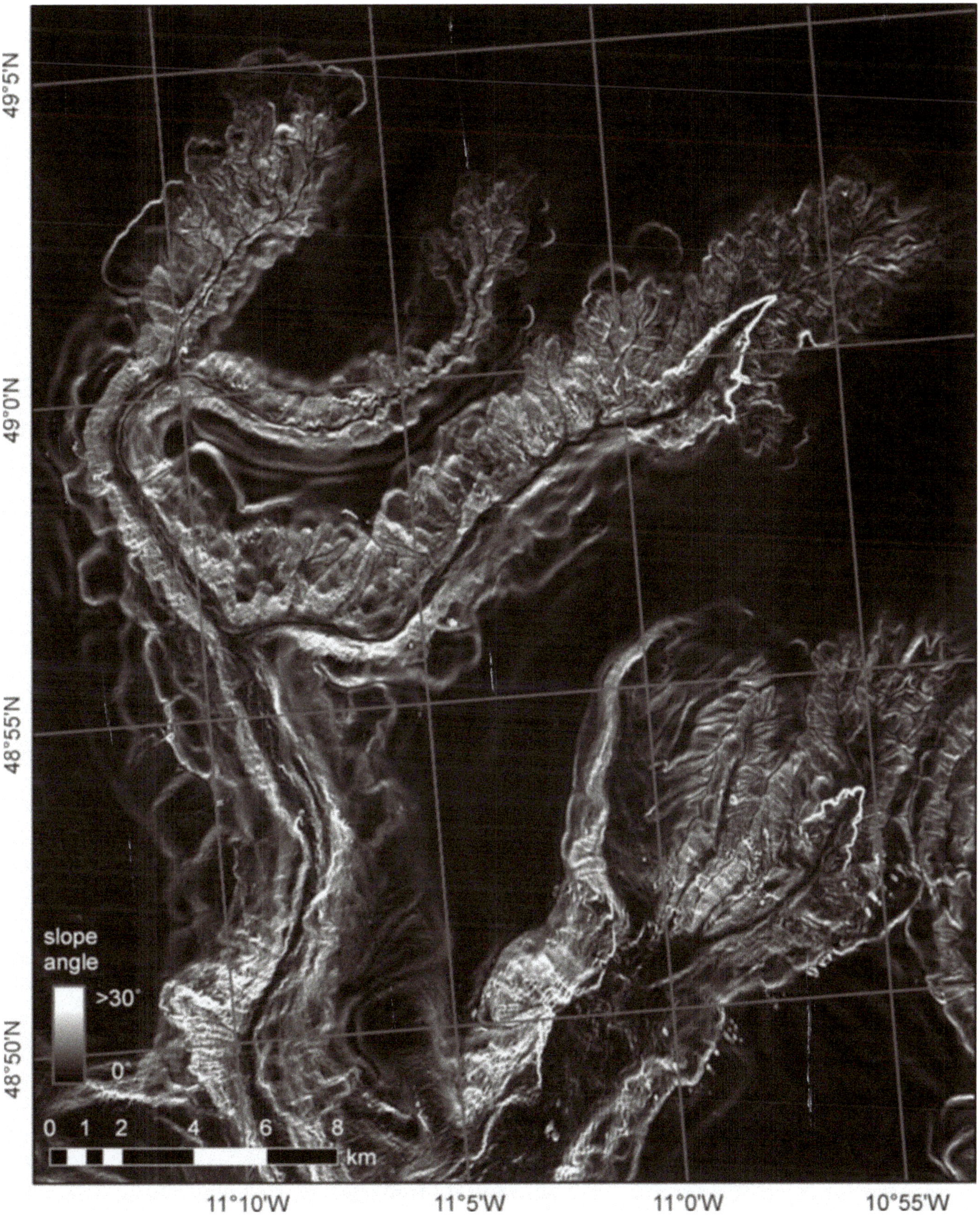

Fig. 7.2 **Canyon detection.** Canyons can be detected based on the increased slope inclination of the canyon walls compared to the surrounding seabed. While the slopes on the "normal" seabed rarely exceed angles steeper than 2°, canyon walls are usually steeper than 5°. In the centre of each canyon, a narrow low angle area (therefore shaded black) is known as the thalweg. It represents the deepest point in canyon cross-sections and is therefore a line that follows the deepest path (= weg) of the valley (= thal). A river flowing down the canyon would follow the thalweg, although in submarine canyons the thalweg is the site of turbidity flows (or flows of dense turbid sediment avalanches). Note, the lobate shapes of the feeder canyons and channels with their head walls clearly displayed on the slope angle image

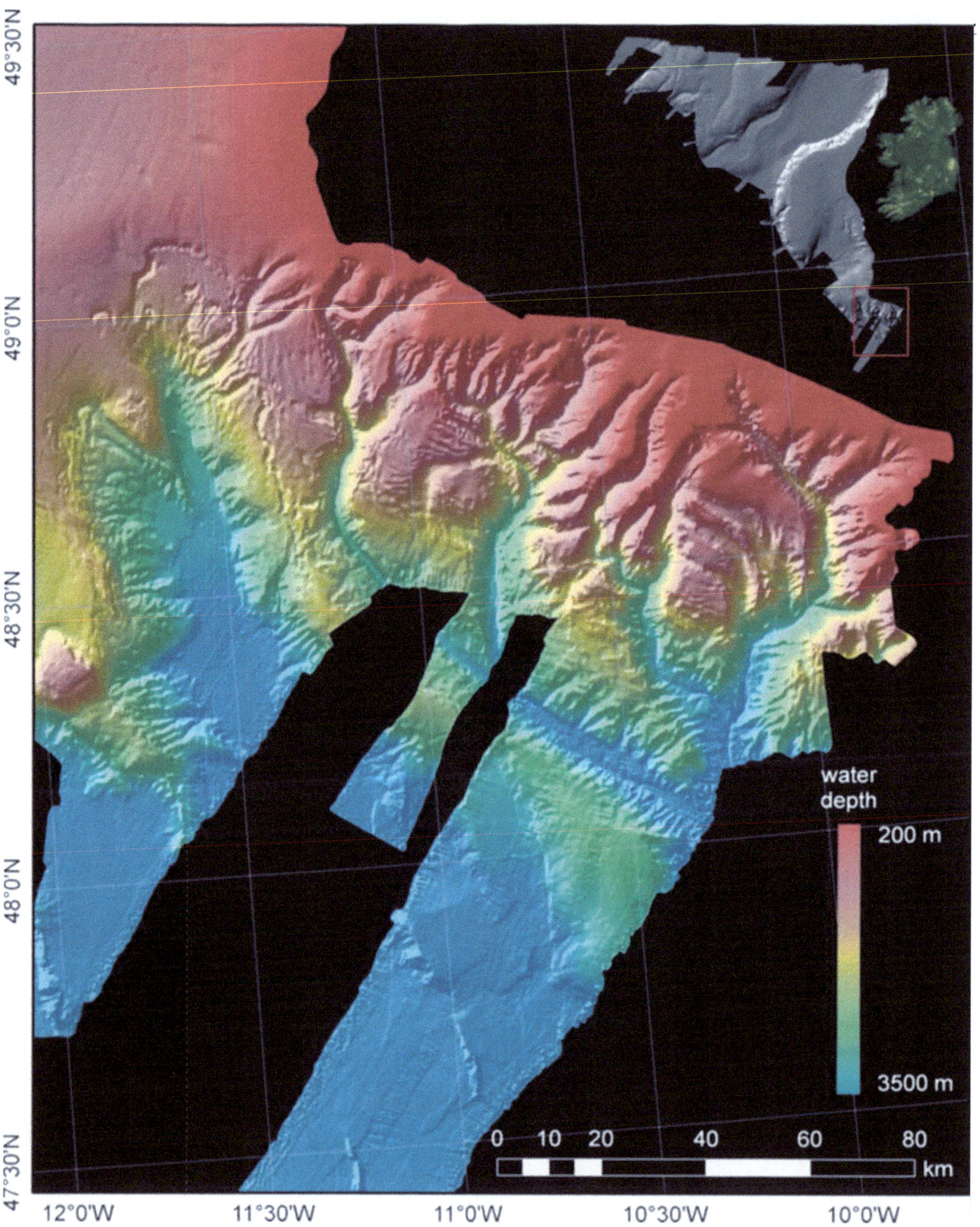

Fig. 7.3 **Whittard Canyon system.** The Whittard Canyon in the SW Approaches is a dendritic canyon system with numerous small feeder canyons that, like river systems on land, successively merge into larger canyons. The Whittard Canyon and other canyon systems along the Celtic continental margin act as fast tracks for the sediments to the deep sea, especially during glacial periods when sea levels are lower and sediment-laden river systems exit at or near the canyon heads

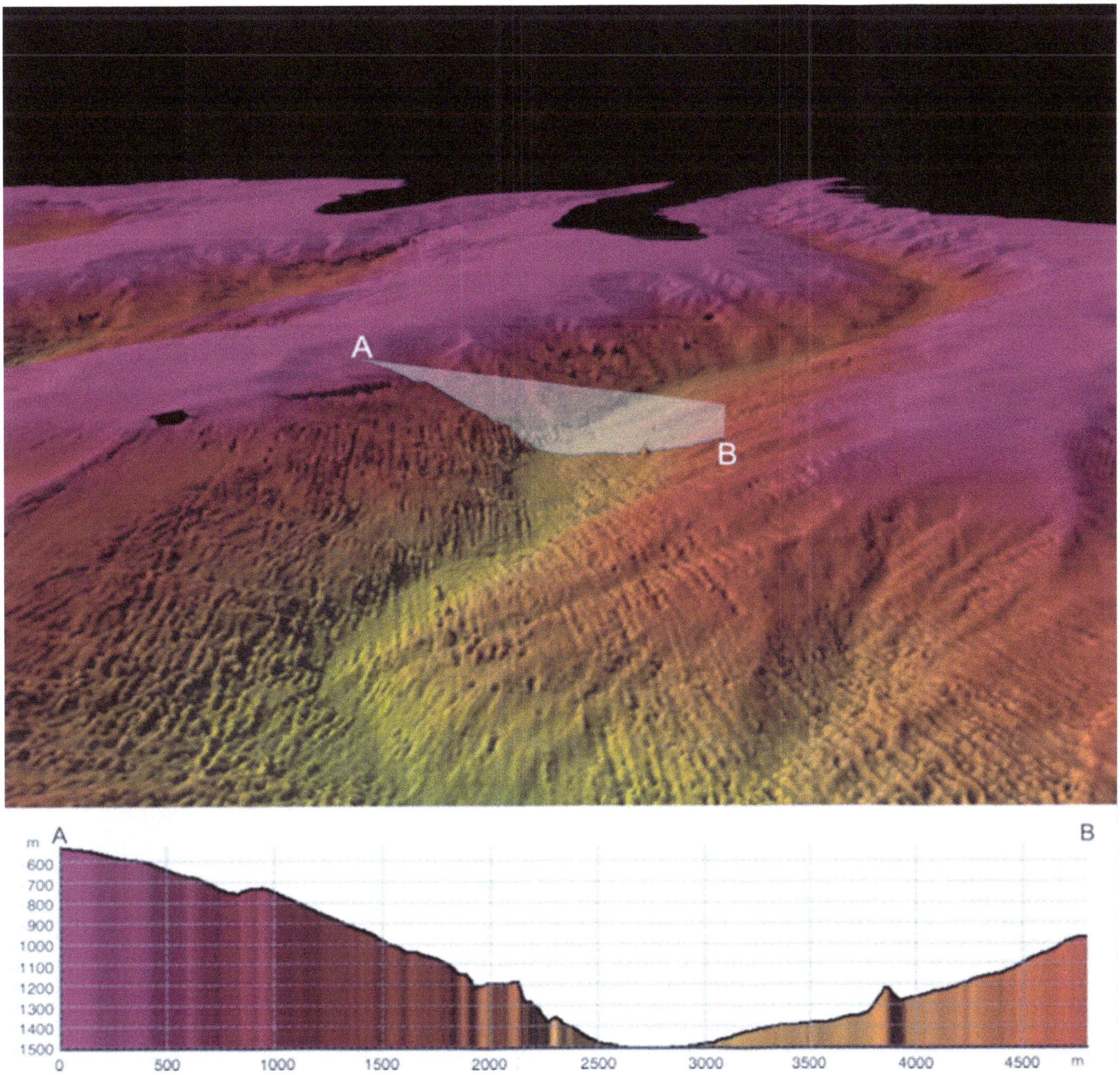

Fig. 7.4 Bathymetric profile across the Whittard Canyon system. The three-dimensional image shows a branch of the upper Whittard Canyon. The line A-B marks the track of the bathymetric profile across the canyon shown in the image below. 3D image and profile are not vertical exaggerated and give therefore an impression of the true steepness of the seabed in this area

Fig. 7.5 **Fauna in the Whittard Canyon.** Corals and crinoids cling to the Whittard Canyon wall at 800 m water depth. The field of view is approximately 2 m across (Image courtesy of the National Oceanography Centre, Southampton, UK (JC10))

Fig. 7.6 Fauna in the Whittard Canyon. Sea fans, anemones and a swat lobster at 4,100 m in the Whittard Canyon. The field of view is approximately 1 m across (Image courtesy of the National Oceanography Centre, Southampton, UK (JC10))

Fig. 7.7 Fauna in the Whittard Canyon. Unusual anemones finding shelter under an over-hang in the channel at the base of the canyon at 2,300 m water depth. The field of view is approximately 50 cm across (Image with courtesy of the National Oceanography Centre, Southampton, UK (JC10))

Fig. 7.8 **Fauna in the Whittard Canyon.** A sea-spider, one of the more intriguing inhabitants of the Whittard Canyon. The field of view is approximately 50 cm across (Image with courtesy of the National Oceanography Centre, Southampton, UK (JC10))

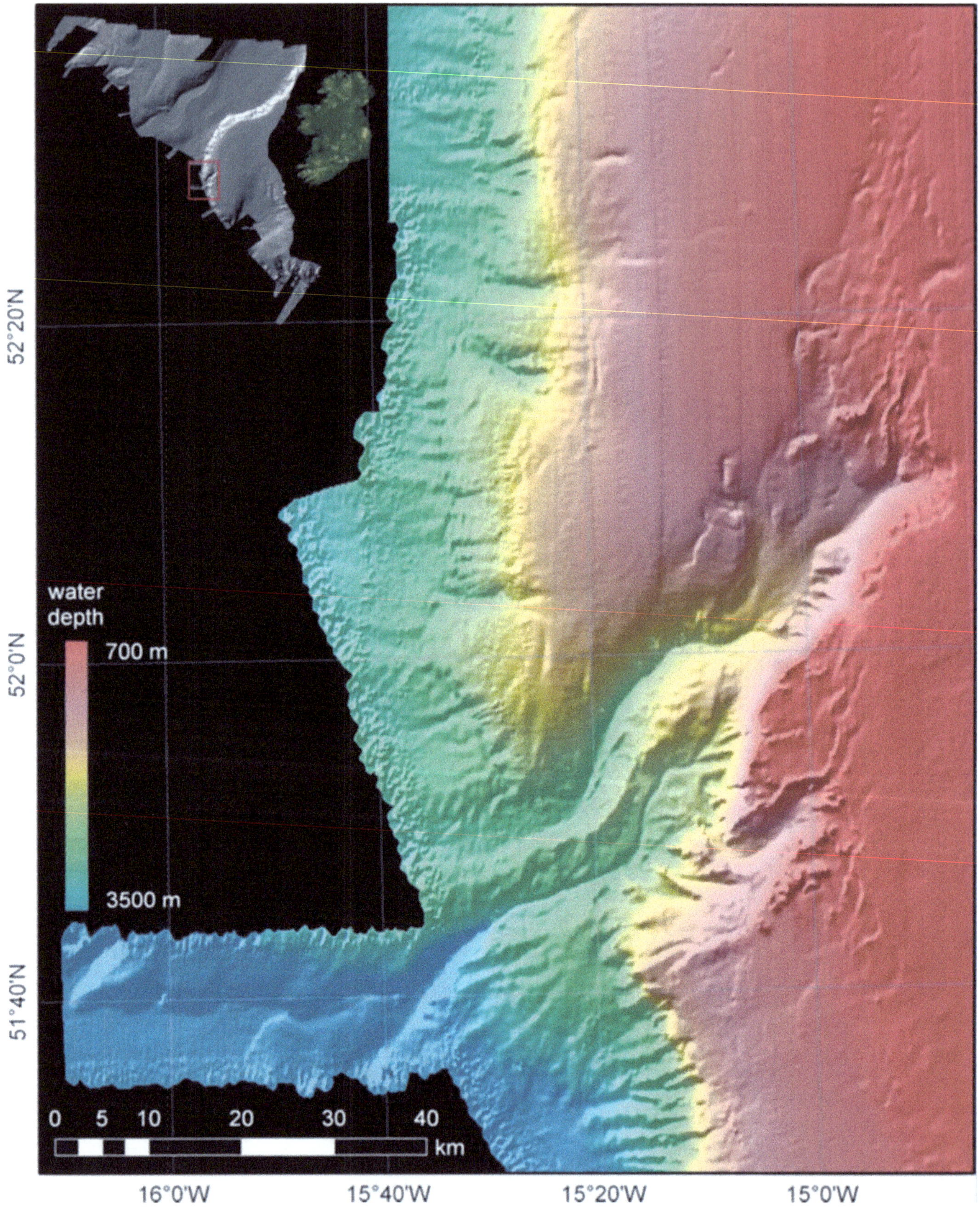

Fig. 7.9 Porcupine Bank Canyon. The Porcupine Bank Canyon cuts deep into the western Porcupine Bank. With its canyon head at approximately 500 m bsl, it is one of the deep seated canyons that cannot be linked to any terrestrial river system

Fig. 7.10 **Fauna in the Porcupine Bank Canyon head.** At the canyon head, consolidated sediments, dead coral frameworks and coral rubble provide hard substrate for various types of corals, anemones, sea urchins, crinoids and sponges. The field of view is approximately 2 m across (Image from the 2009 Offshore Geogenic Reef Mapping Project is provided courtesy of the Department of the Environment, Heritage and Local Government, the Marine Institute and the Geological Survey of Ireland as part of INFOMAR)

Fig. 7.11 **Fauna in the Porcupine Bank Canyon head.** Another common faunal assemblage found at the Porcupine Bank Canyon head consists of cold-water corals, crinoids, ophiuroids and sponges. Species diversity and abundance can be quite high in the canyons. The field of view is approximately 1 m across (Image from the 2009 Offshore Geogenic Reef Mapping Project is provided courtesy of the Department of the Environment, Heritage and Local Government, the Marine Institute and the Geological Survey of Ireland as part of INFOMAR)

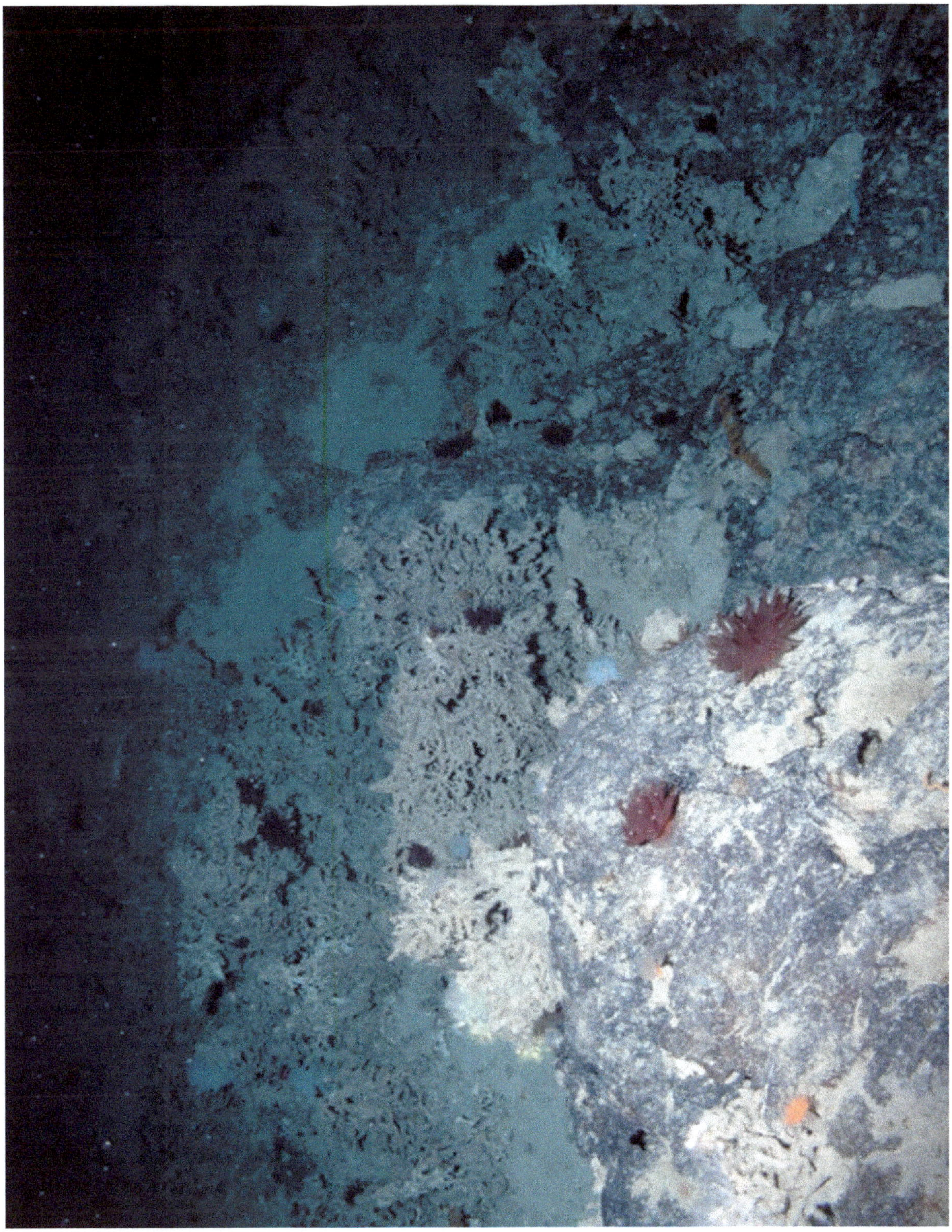

Fig. 7.12 **Fauna in the middle Porcupine Bank Canyon.** Overgrown bedrock and coral thickets at the lower slope of the middle canyon. The field of view is approximately 5 m across (Image from the 2009 Offshore Geogenic Reef Mapping Project is provided courtesy of the Department of the Environment, Heritage and Local Government, the Marine Institute and the Geological Survey of Ireland as part of INFOMAR)

Fig. 7.13 **Fauna in the middle Porcupine Bank Canyon.** The hard substrate provided by the outcropping bedrock supports a high species diversity. Broken-off corals accumulate at the base of the cliff faces. The field of view is approximately 2 m across (Image from the 2009 Offshore Geogenic Reef Mapping Project is provided courtesy of the Department of the Environment, Heritage and Local Government, the Marine Institute and the Geological Survey of Ireland as part of INFOMAR)

Fig. 7.14 **Fauna in the lower Porcupine Bank Canyon.** Consolidated sediments form steep cliffs in the lower part of the Porcupine Bank Canyon which can often be colonised by corals. The field of view is approximately 5 m across (Image from the 2009 Offshore Geogenic Reef Mapping Project is provided courtesy of the Department of the Environment, Heritage and Local Government, the Marine Institute and the Geological Survey of Ireland as part of INFOMAR)

Fig. 7.15 **Fauna in the lower Porcupine Bank Canyon.** Outcropping bedrock also forms cliffs and structured terrain providing hard substrate and shelter for many organisms including fish. The field of view is approximately 2 m across (Image from the 2009 Offshore Geogenic Reef Mapping Project is provided courtesy of the Department of the Environment, Heritage and Local Government, the Marine Institute and the Geological Survey of Ireland as part of INFOMAR)

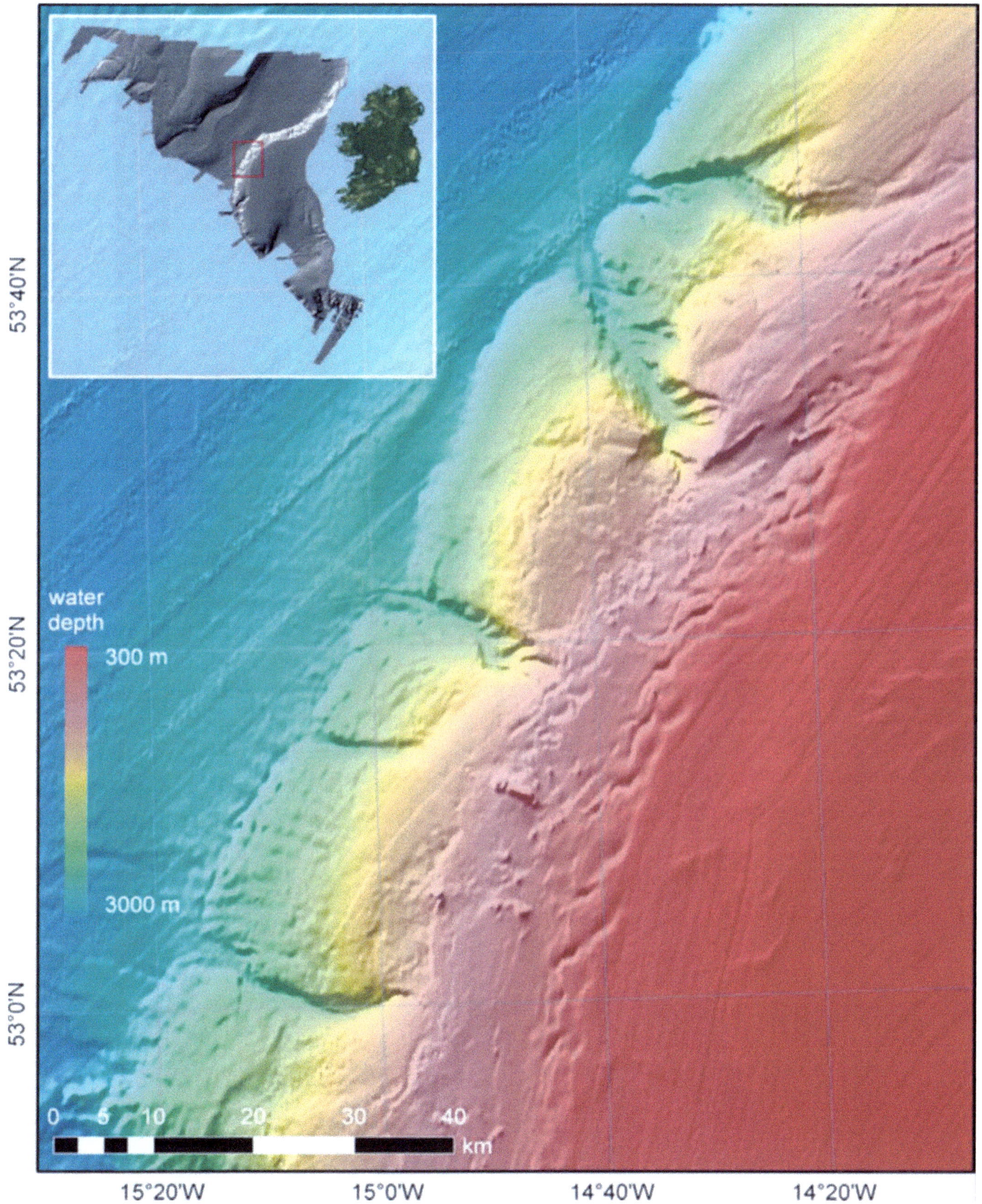

Fig. 7.16 Canyons on the north-west Porcupine Bank. Along the western and northern Porcupine Bank, deep-seated canyons occur every 10–30 km at irregular spaced intervals. Their canyon heads are located below 1,000 m water depth and therefore do not reach up to the shelf break. These canyons are quite a mysterious phenomenon as they neither connect to any river system nor to other sources of sediment supply from the hinterland

Fig. 7.17 **Faunal assemblages in a canyon on the northwest Porcupine Bank.** Different faunal assemblages can be found below 2,000 m bsl in the canyons along the western and northern Porcupine Bank. Here, stalked crinoids and ophiuroids are the most prominent animals. The field of view is approximately 2 m across (Image from the 2009 Offshore Geogenic Reef Mapping Project is provided courtesy of the Department of the Environment, Heritage and Local Government, the Marine Institute and the Geological Survey of Ireland as part of INFOMAR)

Fig. 7.18 **Faunal assemblages in a canyon on the northwest Porcupine Bank.** Bizarre sponges and erosive structures can be found in the deep canyons where hard substrate is also widely available for organisms to settle on. The field of view is approximately 2 m across (Image from the 2009 Offshore Geogenic Reef Mapping Project is provided courtesy of the Department of the Environment, Heritage and Local Government, the Marine Institute and the Geological Survey of Ireland as part of INFOMAR)

Fig. 7.19 **Faunal assemblages in a canyon on the northwest Porcupine Bank.** Brisingid sea stars are a spectacular sight in the deep canyons. The field of view is approximately 1 m across (Image from the 2009 Offshore Geogenic Reef Mapping Project is provided courtesy of the Department of the Environment, Heritage and Local Government, the Marine Institute and the Geological Survey of Ireland as part of INFOMAR)

Fig. 7.20 Rugged terrain in a canyon on the north-west Porcupine Bank. Erosional processes have created rugged terrains in the canyons providing abundant crevices and cavities that provide shelter for deep sea animals. The field of view is approximately 2 m across (Image from the 2009 Offshore Geogenic Reef Mapping Project is provided courtesy of the Department of the Environment, Heritage and Local Government, the Marine Institute and the Geological Survey of Ireland as part of INFOMAR)

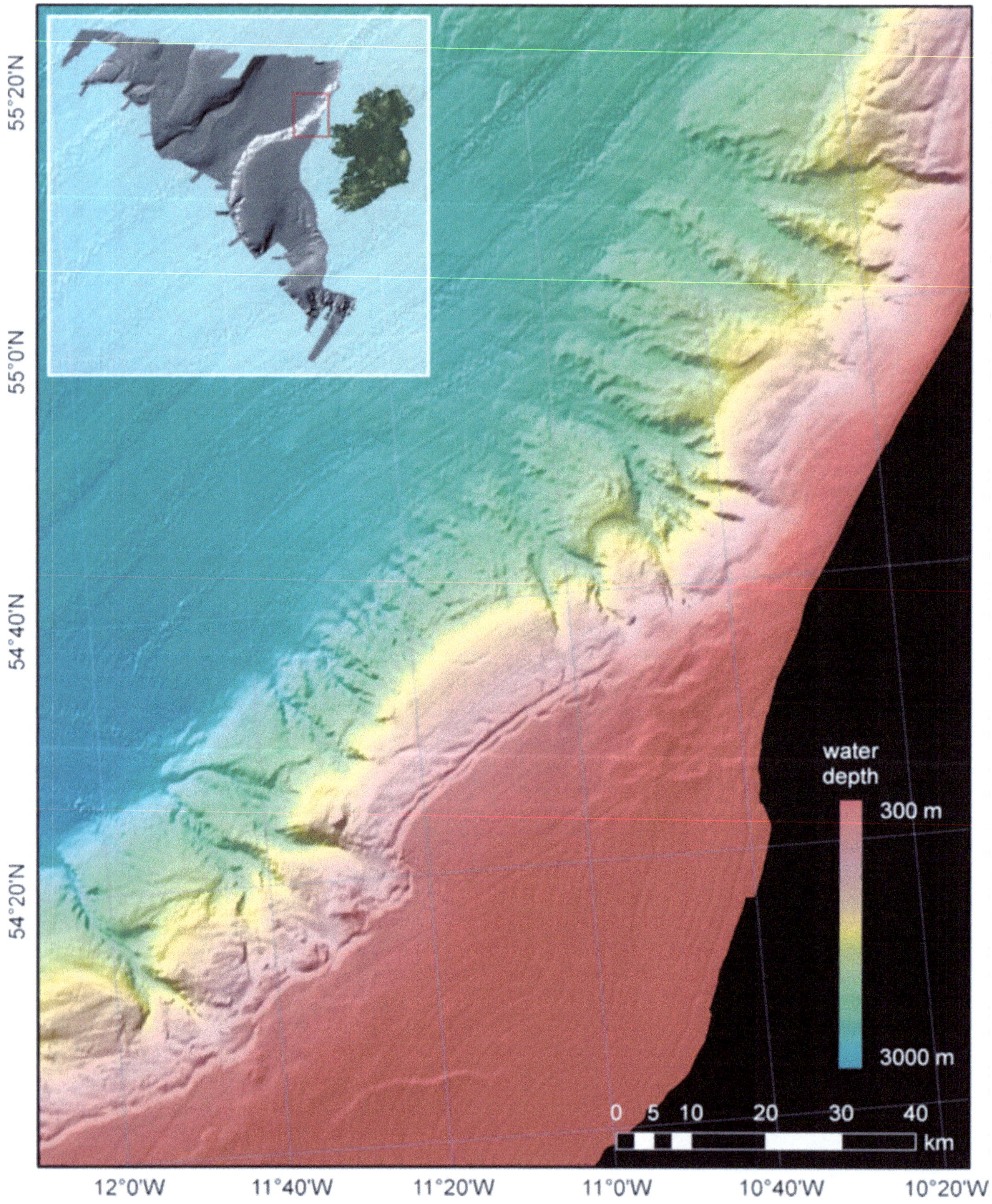

Fig. 7.21 **Canyons along the northernmost Irish continental margin.** Towards the northernmost Irish continental margin, canyons are more closely spaced and the water depth of the canyon heads gradually decreases

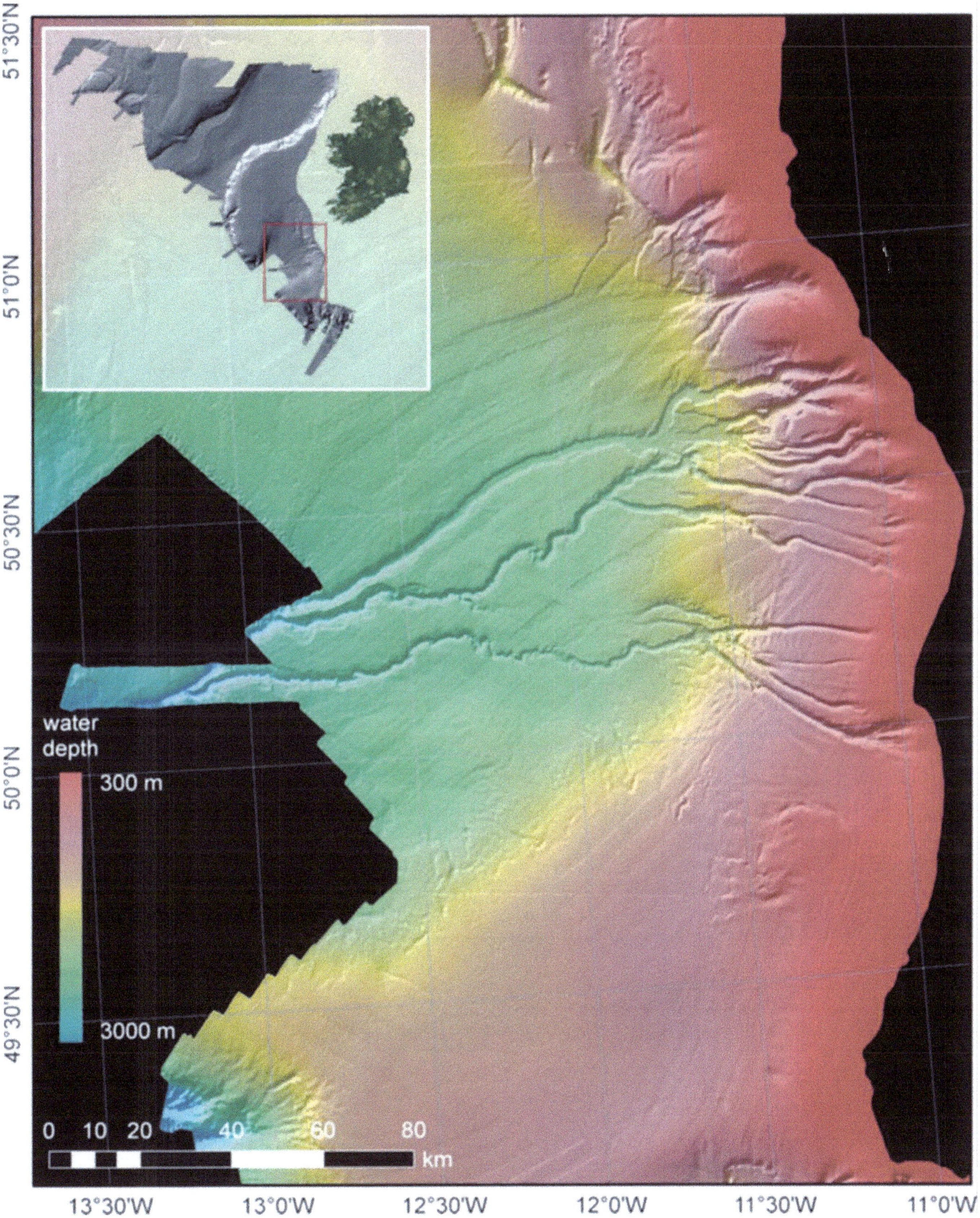

Fig. 7.22 **Gollum Channel.** The Gollum Channel is a large channel system in the Porcupine Seabight. It stretches from the shelf break all the way onto the Porcupine Abyssal Plain beyond the multibeam coverage. It represents the dominant pathway for sediments derived from Ireland during periods of sea level low stand. The Gollum Channel is the largest marine channel system in Europe

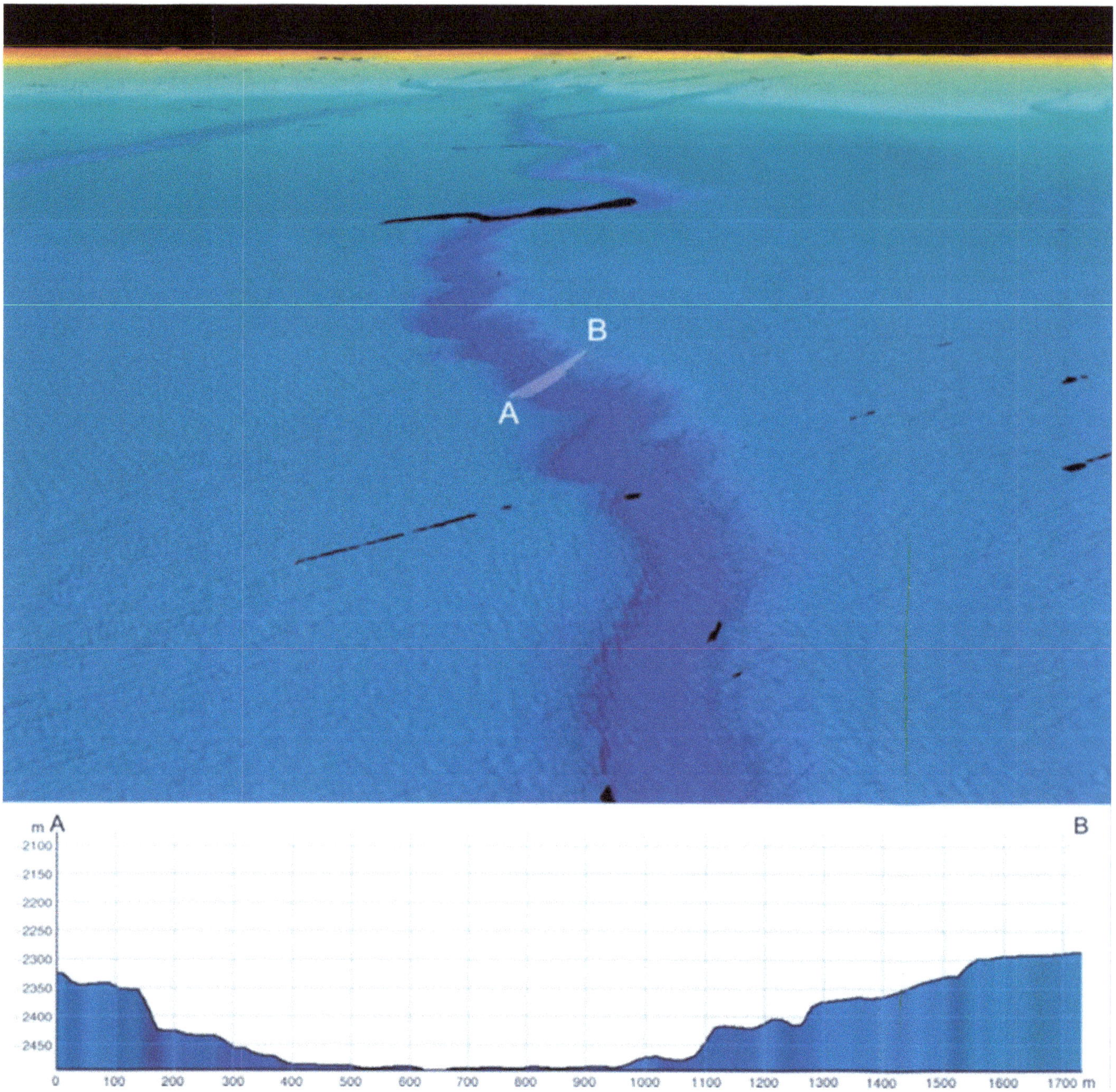

Fig. 7.23 **Bathymetric profile across the Gollum Channel.** The three-dimensional image shows the lower Gollum Channel. The line A-B marks the track of the bathymetric profile across the channel shown in the image below. 3D image and profile are not vertical exaggerated and give therefore an impression of the true steepness of the seabed in this area

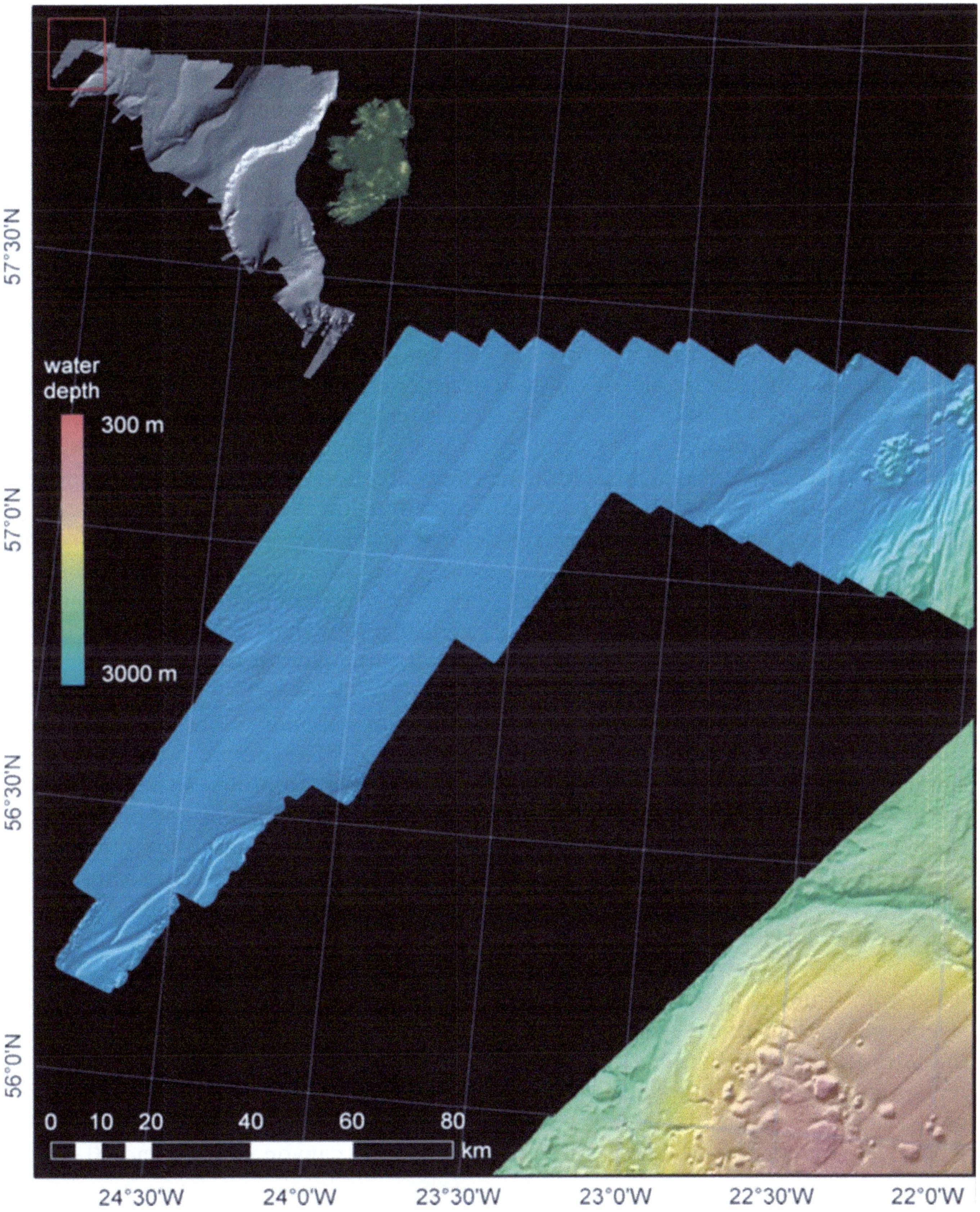

Fig. 7.24 Maury Channel. At the very western tip of multibeam coverage beyond the Hatton Bank and already in the Iceland Basin, parts of the Maury Channel can be seen. The Maury Channel flows southward at the bottom of the Hatton Bank and further to the south at the bottom of the Edoras Bank

been fully solved but may be related to slow scouring by water rigorously pumped up and down the margin by tides with sediment entrained in the canyons at the boundaries between different water-masses.

Sediment transport within canyons is very episodic. Rather than having sediment constantly trickling down the canyons, sediment laden suspensions, so called tubidity currents comparable to snow avalanches on land, flush down the canyons at irregular intervals. These turbidity currents can be triggered by earthquakes or may simply be a result of sediment overloading on steep slopes.

At the canyon mouth, where canyons open onto the deep abyssal plains, sediment fans and depositional lobes often occur. On the Irish seabed, however, these fans are relatively small with strong currents at the base of the continental slope reworking and removing most of the accumulated sediment.

In addition to the submarine canyons, the Irish seabed also contains several submarine channel systems (Fig 7.1). Like submarine canyons, submarine channels are also across-slope pathways for focused transport of suspended sediments carved into the seabed. However, unlike canyons, they occur on gentler slopes, are less deep (tens of metres), less pronounced and are often bordered by levees. In their appearance, some submarine channels resemble rivers meandering down gentle slopes in the terrestrial realm (Fig. 7.22). Channels in the geological record have been the focus of repeated research as their fossil equivalents represent common reservoir rocks for hydrocarbons.

With their steep walls and structured topography, canyons and channels form variable habitats and provide hardgrounds for many organisms such as corals, sponges, hydrozoans and anemones. Across canyons and channels, habitats change quickly resulting in high species diversity.

7.1 Identification

Canyons and channels were identified on the base of hydrological network analyses (HNA). The HNA was performed on a raster dataset of the entire Irish seabed with an approximately cell size of 170 m in the north-south direction. For the HNA, the flow direction for every cell was determined (flowing downslope from higher cells to lower cells) and based on the flow directions, the flow accumulations were calculated. Flow accumulation adds up the numbers of cells that flow into each individual cell. For the identification of the canyons and channels only those cells with an inflow summing up to more than 30 cells were considered. The output was cleaned for artefacts and simplified to represent only the thalwegs of canyons and channels (the thalweg is the line that follows the deepest path of the valley). As canyon walls and the flanks of channels are usually steeper than the surrounding seabed, the outline of canyons and channels can be clearly identified on slope angle maps (Fig. 7.2).

7.2 Distribution

In total, 43 canyon systems and 20 channel systems are located on the Irish continental margin (Fig. 7.1) of which the canyons in the SW Approaches are the most impressive features (Fig. 7.3 & 7.4). These canyons act as pathways for sediments from the English Channel and the Irish Sea to the deep abyssal plain in the Bay of Biscay and form dendritic networks extending for more than 300 km across the Irish, English and French continental margins. Additional large canyon systems occur on the western Porcupine Bank, where a canyon system stretches for more than 90 km from its head at ~600 m water depth to the mouth in the Rockall Trough. Small canyons occur at 30–50 km intervals on the south-western Porcupine Bank (Fig. 7.9), at 15–30 km intervals along the north-western and northern Porcupine Bank (Fig. 7.16) and at 5–15 km intervals further north along the northernmost Irish continental margin (Fig. 7.21). On the Rockall Bank, two deep seated canyons occur on the southern and south-eastern end of the bank.

The longest European channel system is the Gollum Channel in the Porcupine Seabight (Fig. 7.22 & 7.23). The Gollum Channel is at least 300 km long and extends from the shelf break, through the central Porcupine Seabight, through the mouth of the Porcupine Seabight and onto the Porcupine Abyssal Plain. In the Porcupine Seabight, additional shorter feeder channels also occur. Larger channels mark the pathways of the main sediment flows in the Rockall-Hatton Basin and are scoured around seamounts and the Rockall and Hatton Bank. The well developed Maury Channel is partly imaged by the bathymetry data at the very western tip of the Irish seabed (Fig. 7.24).

Bibliography

Bourillet, J.-F., Zaragosi, S., & Mulder, T. (2006). The French Atlantic margin and deep-sea submarine systems. *Geo-Marine Letters, 26*, 311–315.

Clark, J. D., Kenyon, N. H., & Pickering, K. T. (1992). *Quantitative analysis of the geometry of submarine channels: Implications for the classification of submarine fans Geology, 20*, 633–636.

Emery, K. O., & Uchupi, E. (1984). *The geology of the Atlantic Ocean*. New York: Springer.

Kane, I. A., McCaffrey, W. D., & Peakall, J. (2008). Controls on sinuosity evolution within submarine channels. *Geology, 36*, 287–290.

Kenyon, N. H., Belderson, R. H., & Stride, A. H. (1978). Channels, canyons and slump folds on the continental slope between South-West Ireland and Spain. *Oceanologica Acta, 1*, 369–380.

Tudhope, A. W., & Scoffin, T. P. (1995). Processes of sedimentation in Gollum Channel. Porcupine Seabight: Submersible observations and sedimentation analyses. *Transactions of the Royal Society of Edingburgh: Earth Sciences, 86*, 49–55.

Weaver, P. P. E., Wynn, R. B., Kenyon, N. H., & Evans, J. (2000). Continential margin sedimentation, with special reference to the north-east Atlantic margin. *Sedimentology, 47*(1), 239–256.

Wheeler, A. J., Kenyon, N. H., Ivanov, M. K., Beyer, A., Cronin, B. T., McDonnell, A., et al. (2003). Canyon heads and channel architecture of the Gollum Channel, Porcupine Seabight Trough. In J. Mienert & P. Weaver (Eds.), *European margin sediment dynamics: side-scan sonar and seismic images* (pp. 183–186). New York: Springer-Verlag.

Seamounts

8

B. Dorschel, A.J. Wheeler, X. Monteys, and K. Verbruggen

Seamounts are distinct, elevated, geological seabed features, often of volcanic origin. They are at least 1,000 m tall but do not reach the sea surface (Fig. 8.1). The majority of seamounts are conically shaped when young but due to erosion become more irregularly shaped over time with circular to elliptic elongated bases and often flat summit areas. The slopes of seamounts can be quite steep, sometimes up to 60° (Fig. 8.2).

Elevating from the seabed, seamounts protrude into the water column and form obstacles to the flow of water masses. By doing so, seamounts can modify major currents thus amplifying the prevailing currents that pass around them. These interactions often result in complex vortices and current patterns around seamounts that can even be detected in surface currents. One particular phenomenon observed from seamounts is the formation of a so-called Taylor Column: a rotating body of water retained above a seamount. Nutrients can be focused and enriched within Taylor Columns thus locally enhancing food availability, which in turn attracts animals. Seamounts can also interact with internal tides (tides generated along water mass boundaries) which in turn can locally intensify bottom currents up to several tens of centimetres per second. At the seamounts, these amplified currents enhance erosive processes, exposing hard substrate and slowly altering the shape of the seamounts. As a result, seamounts slowly morph into irregular terrains with terraces, canyons, crevices, fissures, pinnacles and hummocks. Other seamount-oceanography interactions include the formation of eddies downstream of seamounts as well as local up- and downwelling of water masses. The upwelling of nutrient rich waters into the photic zone occurs in many areas of the world's oceans and is usually associated with increased biological activity.

For animals, hard substrate, increased food concentrations and increased currents to transport the food to them, make seamounts preferred sites for dwelling. Seamounts are therefore considered biological "hot-spots". But not only this; because of their isolation from the surrounding seabed, seamounts often contain unique faunal assemblages and species found nowhere else in the world. With regard to animal diversity and abundance, the water depth of the summit of the seamount

B. Dorschel (✉) and A.J. Wheeler
Department of Geology & Environmental Research Institute, University College Cork, Cork, Ireland
e-mails: b.dorschel@ucc.ie; a.wheeler@ucc.ie

X. Monteys and K. Verbruggen
Geological Survey of Ireland, Beggars Bush, Haddington Road, Dublin 4, Ireland
e-mails: Xavier.Monteys@gsi.ie; Koen.Verbruggen@gsi.ie

B. Dorschel et al., *Atlas of the Deep-Water Seabed: Ireland*,
DOI 10.1007/978-90-481-9376-9_8, © Springer Science+Business Media B.V. 2010

Fig. 8.1 Seamounts. Three seamounts are located on the Irish seabed. The two seamounts that are located entirely on the Irish seabed occur far in the west, south of Hatton Bank. The Hebrides Terrace Seamount straddles the Irish-British border in the central Rockall Trough

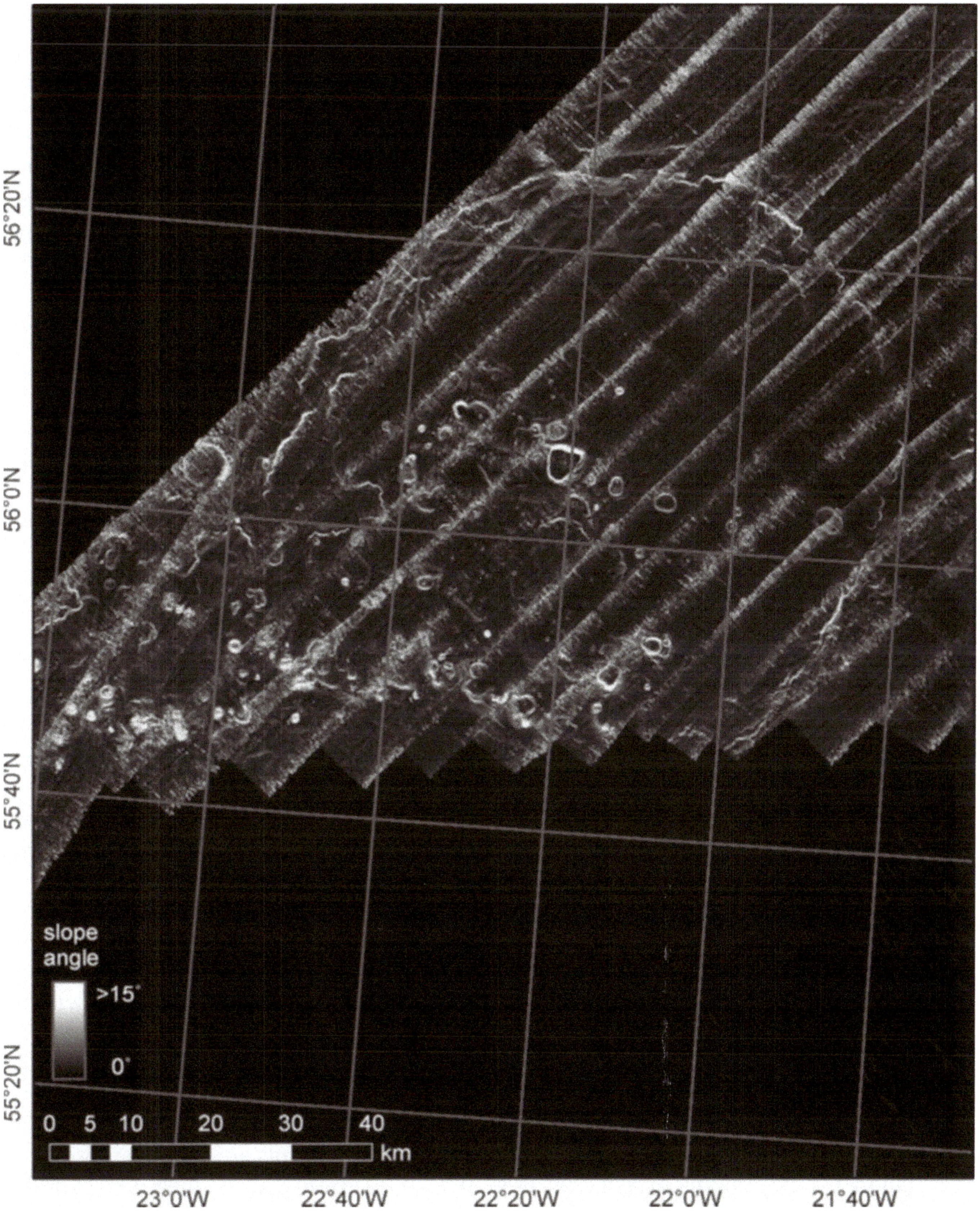

Fig. 8.2 Seamount detection. With slopes steeper than the surrounding seabed, seamounts can best be detected from slope inclination data. On this image the large circle (~85 km diameter) highlighted by diffuse wavy lines of steep slope outline the seamount. The smaller sub-circular (<5 km diameter) features are mounds of unknown origin superimposed on the seamount. The parallel lines running diagonal across the image are artefacts generated in overlapping areas of different survey lines

seems to be an important physical factor. Even seamounts with summit water depths down to 1,000 m bsl are likely to interact with layers of zooplankton migrating vertically in the water column thus providing additional sources of food for the animal communities living on the seamount. The greatest diversity of deep-water corals living on seamounts can be found on mounts with their summit regions shallower than 1,000 m water depth.

This rich biology found at seamounts unfortunately causes the biggest threat for seamount habitats as it attracts fish species targeted by the commercial fishing industry. Catch numbers in the past have proved that seamounts support abundant fisheries but unfortunately many have now been fished out. Deep-water species such as the Orange Roughy with life histories characterised by a late age of sexual maturity, sporadic reproduction and high longevity became quickly overfished. In this sense, the commercial fishery on seamounts can become another example for 'boom and bust' exploitation unless very carefully managed.

Even though seamounts would be considered quite obvious structures on land (with an elevation of 1,281 m, Mount Vesuvius represents a nice terrestrial equivalent), they still represent rather small structures on the seabed. Therefore, the total number of seamounts worldwide can only be estimated. However, a useful physical phenomenon associated with seamounts allows us to make an estimation. As masses of rock, seamounts give rise to small anomalies in the earth's gravitational field pulling water towards themselves. As a result, the sea surface is slightly elevated above seamounts thus indicating the underlying topographic feature. Variations in sea surface heights can be measured very accurately with satellites. According to this indirect approach, the total number of seamounts worldwide has been estimated to 5,000–16,000 features. The large uncertainty results from missing data or resolution issues. Still, about 800 major seamounts are known from the North Atlantic alone occurring mainly along mid-ocean ridges. Others are found at some distance from mid-ocean ridges like, for example, the well known seamounts in the northern (UK) Rockall Trough (Anton Dohrn, Rosemary Bank, Bailley Bank, George Bligh Bank and Lousy Bank). On the Irish seabed, only three seamounts can be found.

8.1 Identification

Typical seamounts (more than 1,000 m high) are quite large geomorphological structures that show up on high resolution DEMs and are easily detected due to their characteristic shape. For most areas of the oceans, these high resolution data are not available. For the Irish seabed, the resolution of the INSS datasets is good enough to map and show seamounts in astonishing detail. Even smaller submarine volcanic structures (mini-seamounts) were identified, mapped and grouped together with the mound features (Chapter 10).

Comparable to the method used for canyons, the outline of seamounts was mapped based on slope inclination (Fig. 8.2). The slopes of seamounts are significantly steeper than the surrounding seabed and therefore show up clearly on images displaying changes in slope (the slope angles). For the

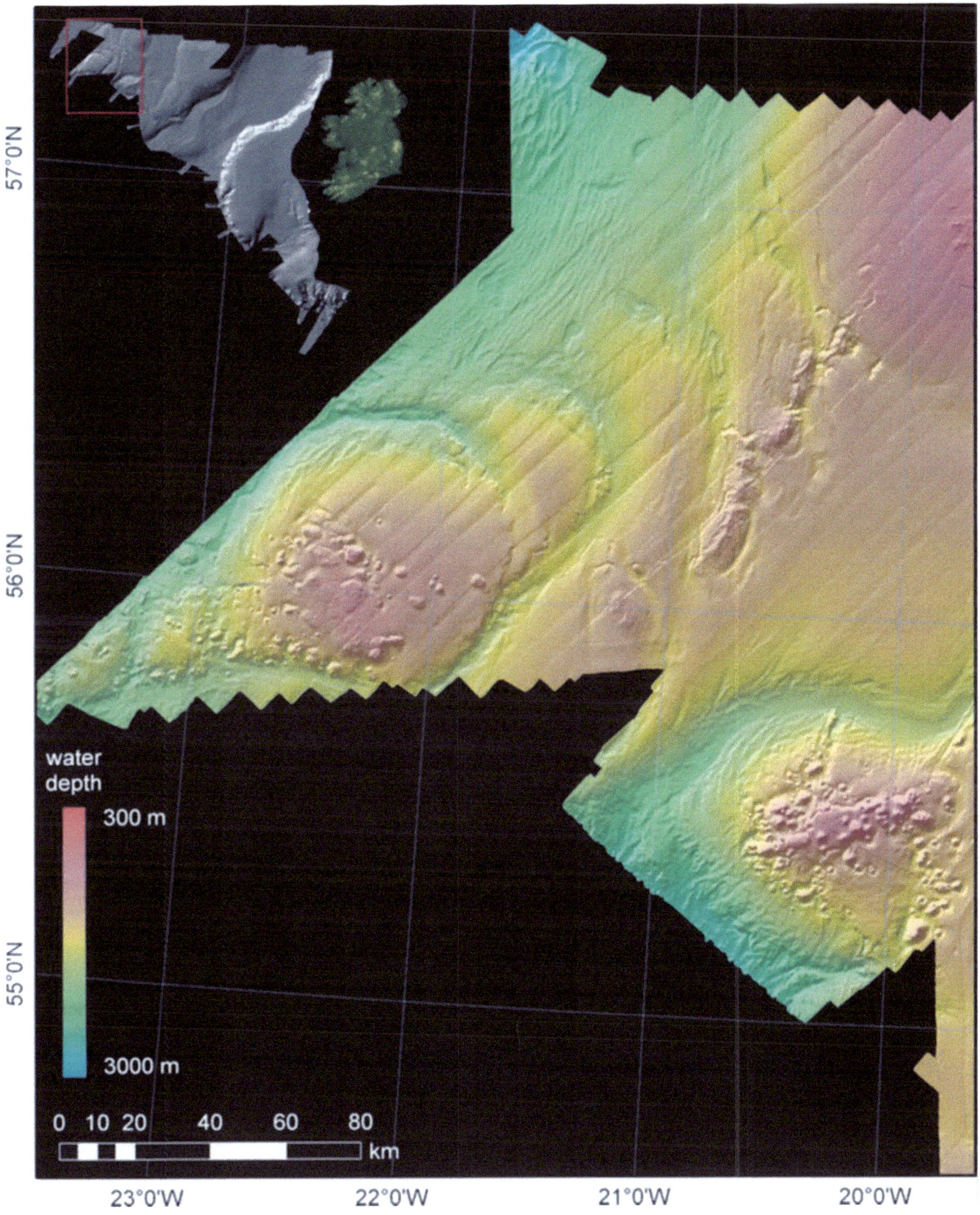

Fig. 8.3 Seamounts south of the Hatton Bank. Two unnamed seamounts are located south of the Hatton Bank. The summits are irregularly shaped and with smaller mounds superimposed. What appears like a third seamount in the top right corner is the southernmost tip of the Hatton Bank

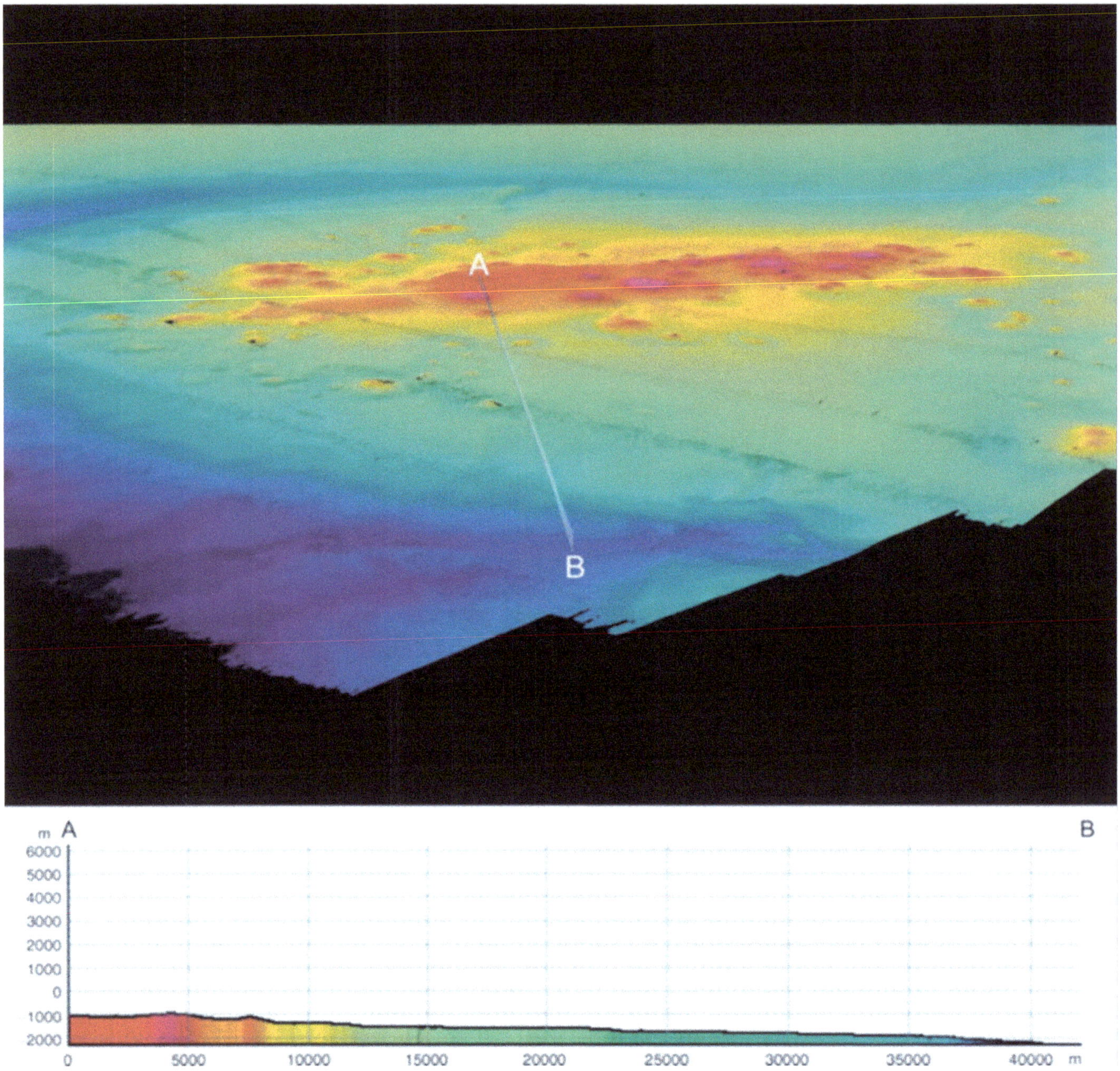

Fig. 8.4 **Bathymetric profile across an unnamed seamount south of the Hatton Bank.** The three-dimensional image shows the southern one of the two seamounts south of Hatton Bank. The line A-B marks the track of the bathymetric profile across the seamount shown in the image below. 3D image and profile are not vertical exaggerated and give therefore an impression of the true steepness of the seabed in this area

Fig. 8.5 **Faunal assemblage on a seamount.** This image, taken from the flank area at ~800 m bsl of an unnamed seamount west of Rockall Bank, shows the wide abundance of hard substrate. Corals, sponges, serpulid worms, crabs and many other species take advantage of the exposed rocks on which they can settle. The field of view is approximately 2 m across (Image courtesy of MARUM – Center for Marine Environmental Sciences, University of Bremen)

Fig. 8.6 **Local morphology on a seamount.** Hard substrate is widely available on the seamount but not all is used as settlements by organisms. Exposed rocks as well as accumulations of pebble and boulder sized rocks are indicative of strong bottom currents and the generally erosive environment at the seamount. At the summit of this seamount, soft sediments were restricted to sheltered areas. The field of view is approximately 5 m across (Image courtesy of MARUM – Center for Marine Environmental Sciences, University of Bremen)

Fig. 8.7 **Fauna and morphology at a seamount.** At seamounts the concentration of particles in the water can be quite high and almost blizzard-like in appearance. The high concentration of particles, which including food particles for filter feeding organisms, supports the rich biology at seamounts. The field of view is approximately 2 m across (Image courtesy of MARUM – Center for Marine Environmental Sciences, University of Bremen)

Fig. 8.8 **A close up image of a seamount faunal assemblage.** This close up image shows corals, sponges, serpulid worms etc. growing in close proximity. It is apparent that most of the species live in the sheltered areas at the edges of the slab of rock. On top of the slab, the density of organisms is significantly lower and well protected barnacles are most common. The field of view is approximately 50 cm across (Image courtesy of MARUM – Center for Marine Environmental Sciences, University of Bremen)

detection of seamounts (large and small), a critical slope angle of 6° was applied as cut-off value. However, a distinction is made here between seamounts (exceeding 1,000 m height) and smaller features called mounds. A special type of mound formed from coral reefs is dealt with separately under the section on cold-water coral carbonate mounds (Chapter 10).

8.2 Distribution

Three seamounts are located on the Irish seabed. Two unnamed seamounts are located south of the Hatton Bank (Fig. 8.3), and the Hebrides Terrace Seamount is located on the Irish-UK border in the central Rockall Trough. These true seamounts measure between 36 and 110 km across and 1,000–1,500 m from base to summit. In addition to the seamounts, one cluster of groundtruthed mini-seamounts is located at the foot of the northwest corner of the Porcupine Bank. There, 51 volcanic mounds (0.5–3.5 km across; 60–460 m high) occur in water depths between 2,900 and 2,000 m. It is likely that many more mini-seamounts (mound features of volcanic origin) are located on the Irish seabed but a clear distinction is not possible due to a lack of groundtruthing data. Chapter 10 'Mound Features and Coral Carbonate Mounds' deals specifically with the distribution of these smaller mound features that can be roughly divided into volcanic features, erosional features (e.g. outcropping bedrock) and biogenic features (coral carbonate mounds). Towards the western extent of the Irish seabed, groundtruthing information is very sparse and the origin of many of the mound features in this area is impossible to determine.

Bibliography

Clark, M. R., Tittensor, D., Rogers, A. D., Brewin, P., Schlacher, T., Rowden, A., et al. (2006). *Seamounts, deep-sea corals and fisheries: Vulnerability of deep-sea corals to fishing on seamounts beyond areas of national jurisdiction*. Cambridge, UK: UNEP-WCMC.

Genin, A., Dayton, P. K., Lonsdale, P. F., & Spiess, F. N. (1986). Corals on seamount peaks provide evidence of current acceleration over deep-sea topography. *Nature, 322*, 59–61.

Glover, L. K., & Earle, S. A. (Eds.). (2004). *Defying ocean's end* (pp. 45–70). Washington, DC: Island Press.

Gubbay, S. (2003). *Seamounts of the North-East Atlantic*. WWF Germany: OASIS.

Pitcher, T. J., Morato, T., Hart, P. J. B., Clark, M. R., Haggan, N., & Santos, R. S. (2007). *Seamounts: Ecology, fisheries & conservation*. Oxford: Blackwell. 552 pp.

Roberts, D. G., Hogg, N. G., Bishop, D. G., & Flewellen, C. G. (1974). Sediment distribution around moated seamounts in the Rockal Trough. *Deep Sea Research and Oceanographic Abstracts, 21*, 175–184.

Roberts, C. M. (2002). Deep impact: The rising toll of fishing in the deep sea. *Trends in Ecology and Evolution, 17*(5), 242–245.

Rogers, A. D. (1994). The biology of seamounts. *Advances in Marine Biology, 30*, 305–350.

Stone, G. S., Madin, L. P., Stocks, K., Hovermale, G., Hoagland, P., Schumacher, M., et al. (2004). Seamount biodiversity, exploitation and conservation. In L. K. Glover & S. A. Earle (Eds.), *Defying ocean's end* (pp. 45–70). Washington: Island Press.

Weaver, P. P. E., Billett, D. S. M., Boetius, A., Danovaro, R., Freiwald, A., & Sibuet, M. (2004). Hotspot ecosystem research on Europe's deep-ocean margins. *Oceanography, 17*(4), 132–143.

Escarpments

9

B. Dorschel, A.J. Wheeler, X. Monteys, and K. Verbruggen

In a geological/geographical context, the term escarpment is defined as a long cliff or steep slope that has resulted from erosional processes or faulting. The specific reasons why escarpments form on the seabed can be many fold (although faulting is no longer an active process on the Irish margin). Escarpments can form as a result of strong, slope-parallel bottom currents, or as the scars left following submarine (land)slides. In addition, they can be associated with morphological features such as canyons and seamounts. Regardless of the formation processes, escarpments are the results of erosive processes active on the continental slope.

In contrast to common perceptions, the deep oceans are not quiet, steady-state environments but are in fact quite dynamic with intense bottom currents common in many areas, especially along the continental slopes. The interaction of meridional currents (currents that flow polewards) with the slopes can result in seabed hugging "contour currents". As the oceans are composed of different layers of water with different densities, temperatures and salinities (called water masses) that flow in different directions at different depth levels, contour currents can be confined to distinct depth intervals. This limits erosion of the seabed to only certain depths generating long slope-parallel erosion features (Fig. 9.1).

Another phenomenon that causes erosion of the seabed at distinct depth intervals is the focusing of internal tides. At the boundaries between water masses with significantly different densities, internal tides (similar to tides on the sea surface) can focus energy and locally generated strong tidal bottom currents. If the density changes rapidly between two water masses, this results in a steep density gradient called a pychnocline. At pychnoclines, frictional drag between two differentially moving water masses can form waves (just like the wind blowing across the sea forms waves at the sea–air interface). These are called internal waves and they can locally amplify bottom current speeds to almost 1 m per second during peak flows. These tidal bottom currents have already been recorded in many places along the Irish continental slope using instrumented bottom lander systems and current meter moorings.

Erosive bottom currents can exceed tens of centimetres per second with peak flows often in excess of 50 cm s^{-1} (or ~1 knot). At these

B. Dorschel (✉) and A.J. Wheeler
Department of Geology & Environmental Research Institute, University College Cork, Cork, Ireland
e-mails: b.dorschel@ucc.ie; a.wheeler@ucc.ie

X. Monteys and K. Verbruggen
Geological Survey of Ireland, Beggars Bush, Haddington Road, Dublin 4, Ireland
e-mails: Xavier.Monteys@gsi.ie; Koen.Verbruggen@gsi.ie

B. Dorschel et al., *Atlas of the Deep-Water Seabed: Ireland*,
DOI 10.1007/978-90-481-9376-9_9, © Springer Science+Business Media B.V. 2010

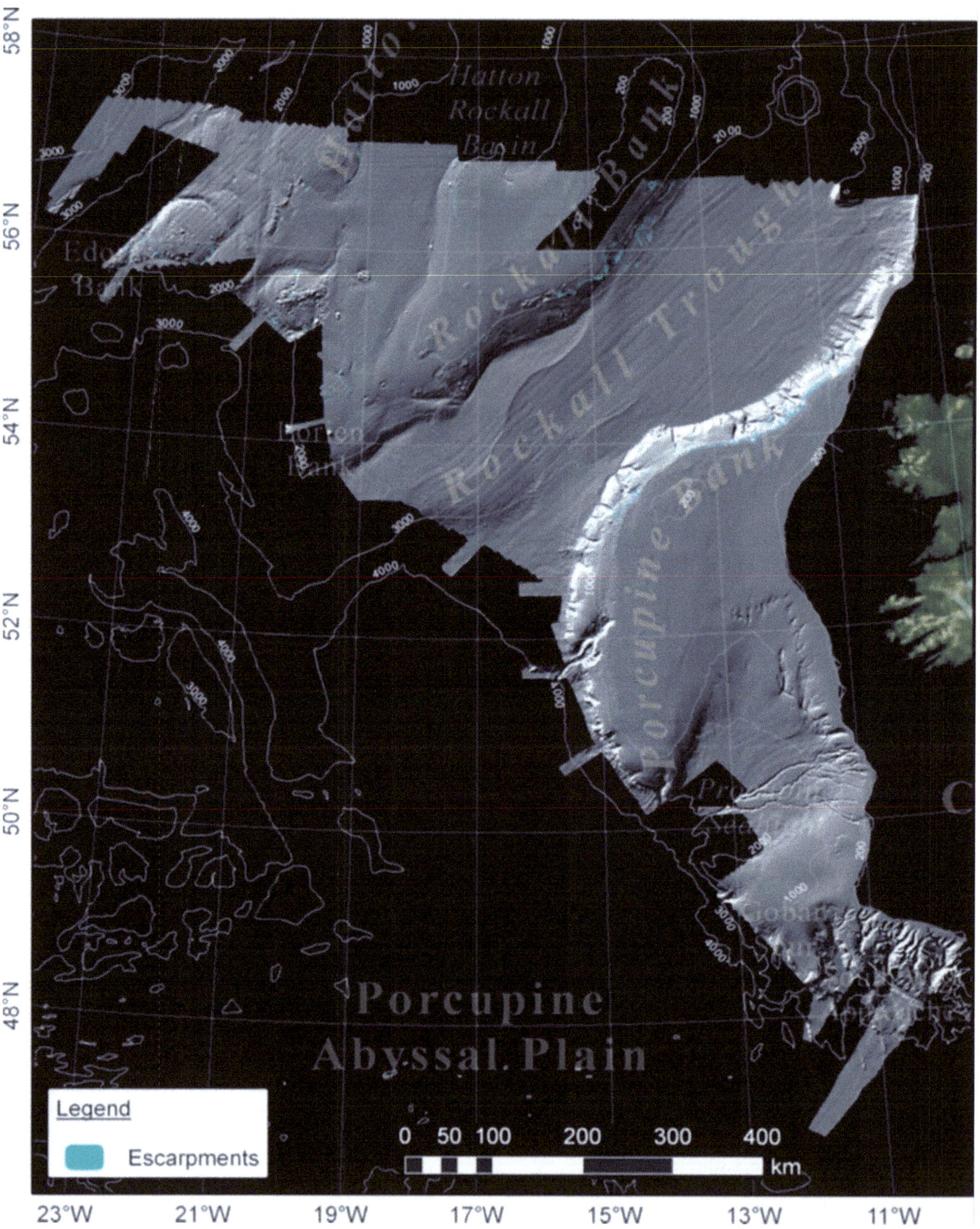

Fig. 9.1 **Escarpments.** On the Irish seabed, most escarpments occur along the upper eastern and western margins of the Rockall Trough. Additional escarpments are associated with canyons and seamounts. Escarpments are always the result of erosional processes. These processes can be erosive bottom currents or slumping and sliding events

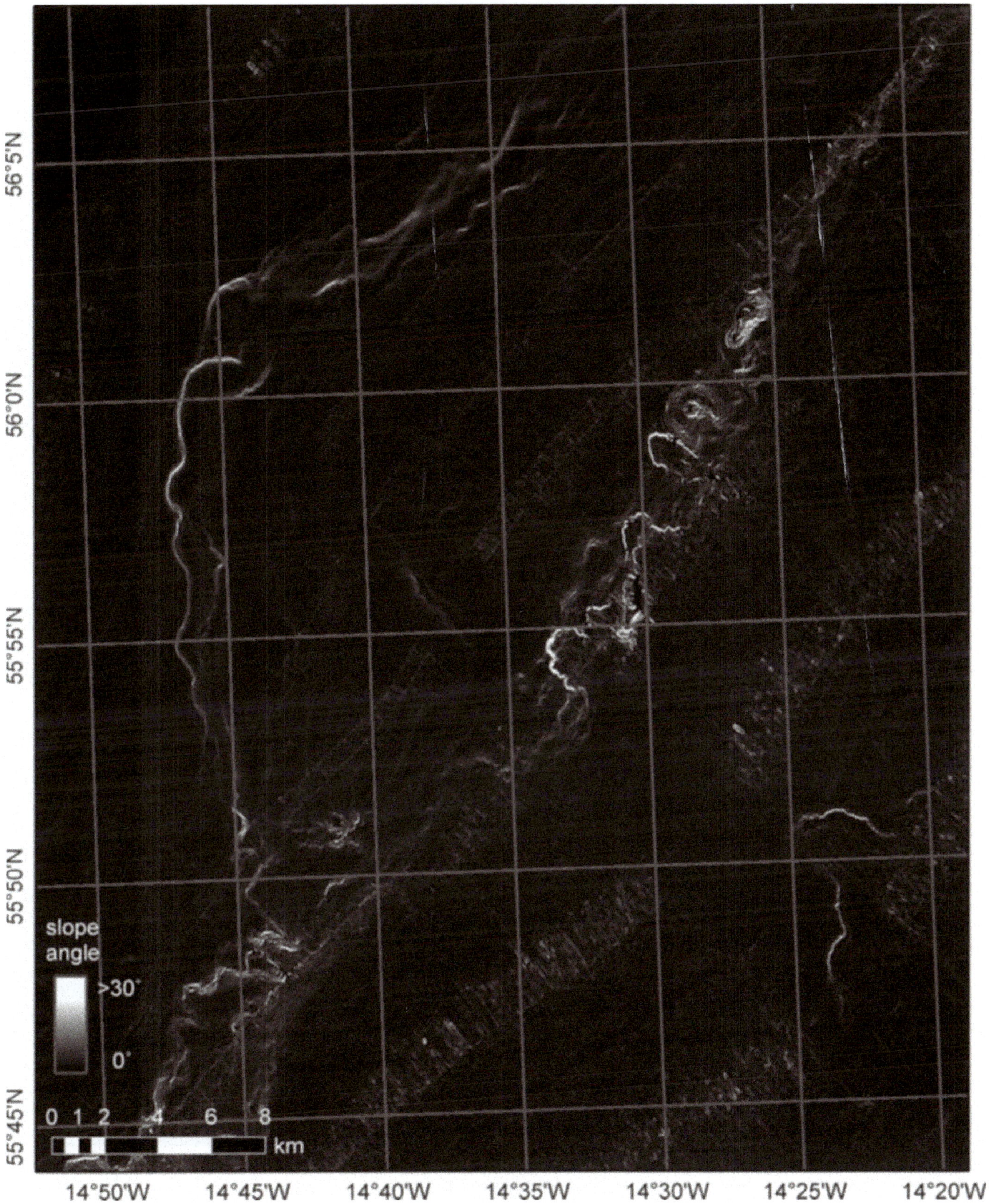

Fig. 9.2 Escarpment detection. Escarpments were detected based on slope maps where areas of steep inclination show up in light shades. The image covers the eastern slope of the Rockall Bank towards the northern extension of the Irish seabed. Escarpments show up well on the slope inclination map and bifurcating lines of escarpments are clearly visible

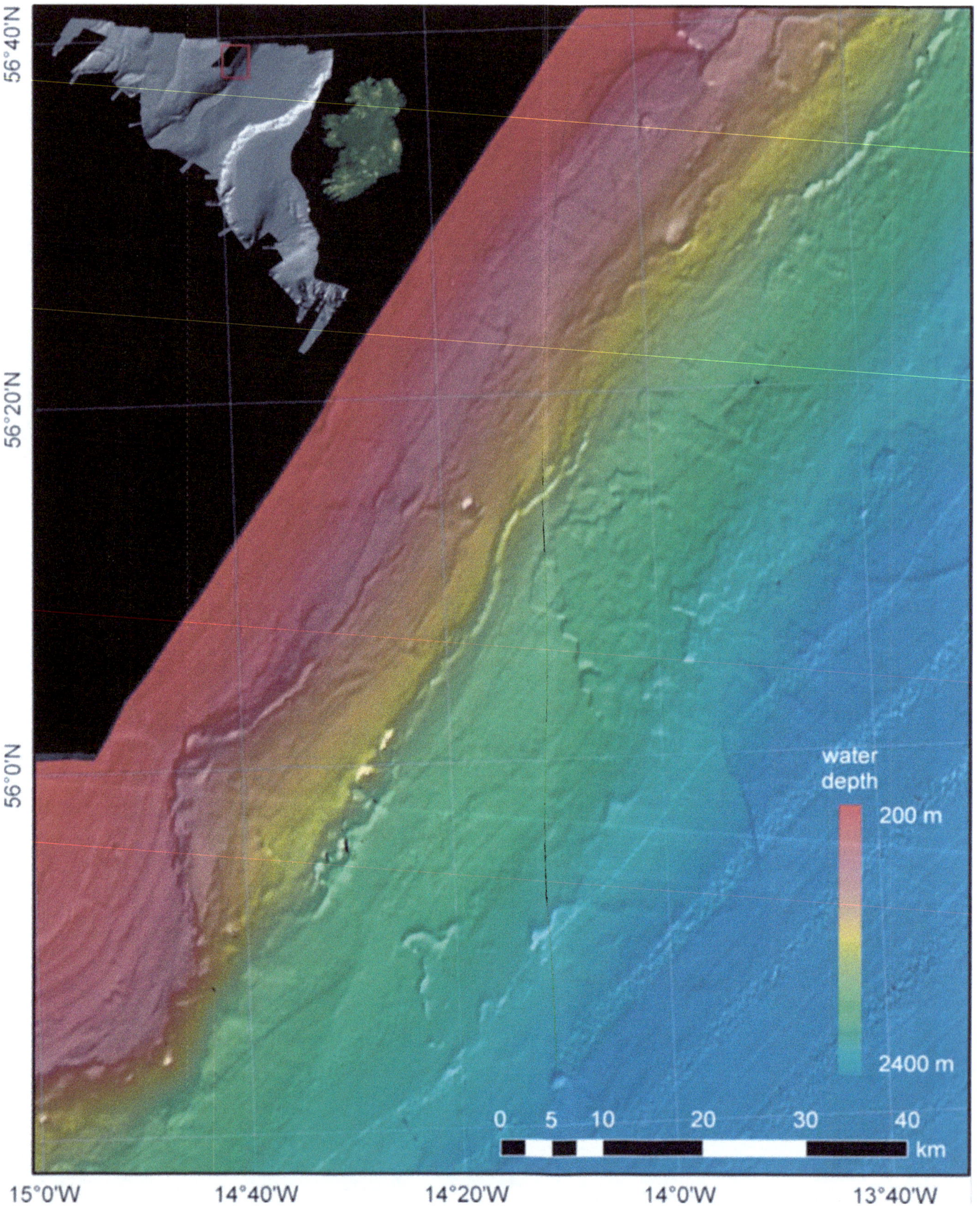

Fig. 9.3 Escarpments on the eastern Rockall Bank. Along the Rockall Bank, several pronounced escarpments occur. Two more or less parallel lines of escarpments can be found in ~650 m to ~1,500 m water depth. Further down the slope, additional smaller escarpments are located. While the upper lines of escarpments are the results of erosive, slope-parallel bottom currents, the escarpments deep on the slope are probably headwalls of sediment slumps and slides

high flow speeds, the energy of the contour current is sufficient to re-suspend sediments, easily moving sands and even gravel. Where strong bottom currents remain focused over long periods of time, soft or incompetent sediments become eroded leaving the underlying cemented or competent sediment exposed. The erosion of unconsolidated sediment can also undercut more consolidated, lithified and partly-lithified sediment layers. At some stage, the overhanging parts of the consolidated units become unstable and collapse to form cliffs. At the bottom of the escarpments along the Irish continental slope, slabs of such collapsed lithified sediments can often be found in chaotic piles.

The depth interval at which erosive currents occur is dictated by regional hydrography. As an example, water that leaves the Mediterranean Sea, so-called Mediterranean Outflow Water (MOW), forms a strong erosive northward directed contour current flowing along the SW Approaches, the Porcupine Seabight and the Porcupine Bank. The MOW, quite salty and denser than the Atlantic surface water, gradually sinks and can be found at 400–900 m water depth where many escarpments occur.

Submarine slumps and slides can also leave escarpments. These are known as slide headwalls and can be found along steep areas of the slope and especially in the walls and headwalls of canyons and along the steep slopes of seamounts. Although not always easy to distinguish, only stand-alone escarpments (those which did not co-exist with other features) were mapped for the atlas. Escarpments associated with canyons and channels or seamounts were not individually mapped for reasons of visual clarity even though they form a tiered seabed with very steep slopes (Fig. 9.3).

9.1 Identification

Escarpments were identified with cell-by-cell slope analyses of 25 m resolution bathymetry raster data sets. The cut-off value for escarpments was a critical slope angle of 11° or an inclination of 19.4% respectively. The areas with slopes steeper than 11° were mapped and the output was cleaned for artefacts to include escarpments that were not coincident with other features (Fig. 9.2).

9.2 Distribution

Escarpments occur throughout the Irish seabed although mainly on the upper continental slope (Fig. 9.1) and are very pronounced on the Rockall Bank (Fig. 9.3 & 9.4) and the Porcupine Bank (Fig. 9.12). Along the western, north-western and northern sides of the Porcupine Bank, escarpments occur between 650 and 450 m water depth and cover a length of ~120 km marking the change from the rather flat top of the Porcupine Bank to the steeper slope. The escarpments along the eastern Rockall Bank are located in deeper water depths than those on the west and occur mainly below 750 m water depth with an even deeper second zone of escarpments below 1,400 m water depth towards the northern extension of the Irish seabed on the eastern Rockall Bank (Fig. 9.3). On the western Rockall Bank, escarpments occur towards the northern end of the Irish seabed in water depth below 800 m. Additional smaller escarpments also occur along the west and south Rockall Bank and the Hatton Bank.

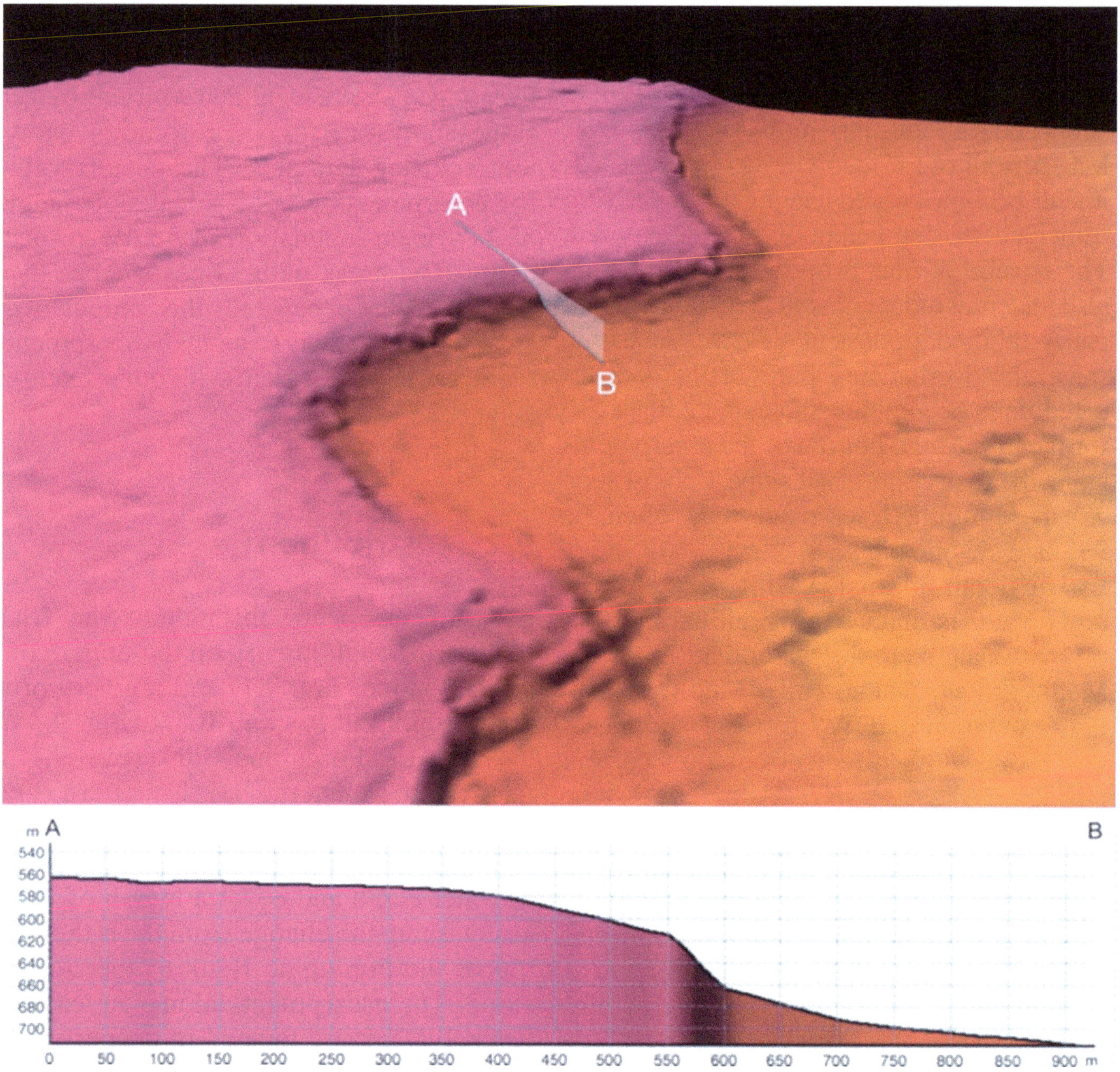

Fig. 9.4 Bathymetric profile across an escarpment on the upper eastern Rockall Bank. The three-dimensional image shows an escarpment along the upper eastern Rockall Bank. The line A-B marks the track of the bathymetric profile across the escarpment shown in the image below. 3D image and profile are not vertical exaggerated and give therefore an impression of the true steepness of the seabed in this area

Fig. 9.5 **A cliff face on an escarpment on the eastern Rockall Bank.** The image taken with a downward-looking camera shows one of the vertical cliff faces at the lower line of escarpments shown in the previous figure from the eastern Rockall Bank (Fig. 9.4). At the escarpments, the seafloor can drop in several vertical steps for several hundreds of metres. The field of view is approximately 2 m across. (Image from the 2009 Offshore Geogenic Reef Mapping Project is provided courtesy of the Department of the Environment, Heritage and Local Government, the Marine Institute and the Geological Survey of Ireland as part of INFOMAR)

Fig. 9.6 **Vertical cliffs on an escarpment on the eastern Rockall Bank.** The field of view is approximately 5 m across. (Image from the 2009 Offshore Geogenic Reef Mapping Project is provided courtesy of the Department of the Environment, Heritage and Local Government, the Marine Institute and the Geological Survey of Ireland as part of INFOMAR)

Fig. 9.7 Fauna on the upper slope of an escarpment on the eastern Rockall Bank. Corals, brisingid sea stars, brittle stars, sponges and many more organisms dwell on the cliff faces of the escarpments. The field of view is approximately 2 m across. (Image from the 2009 Offshore Geogenic Reef Mapping Project is provided courtesy of the Department of the Environment, Heritage and Local Government, the Marine Institute and the Geological Survey of Ireland as part of INFOMAR)

Fig. 9.8 **Fauna on the upper slope of an escarpment on the eastern Rockall Bank.** Brisingid sea stars climb on corals to reach higher into the currents sweeping food along the escarpments. The field of view is approximately 5 m across. (Image from the 2009 Offshore Geogenic Reef Mapping Project is provided courtesy of the Department of the Environment, Heritage and Local Government, the Marine Institute and the Geological Survey of Ireland as part of INFOMAR)

Fig. 9.9 Biogenic rubble at the foot of an escarpment on the eastern Rockall Bank. When organisms die off, get bio-eroded and become disconnected from the vertical walls, they sink to the foot of the escarpments where they accumulate as biogenic (mainly coral) rubble. The field of view is approximately 2 m across. (Image from the 2009 Offshore Geogenic Reef Mapping Project is provided courtesy of the Department of the Environment, Heritage and Local Government, the Marine Institute and the Geological Survey of Ireland as part of INFOMAR)

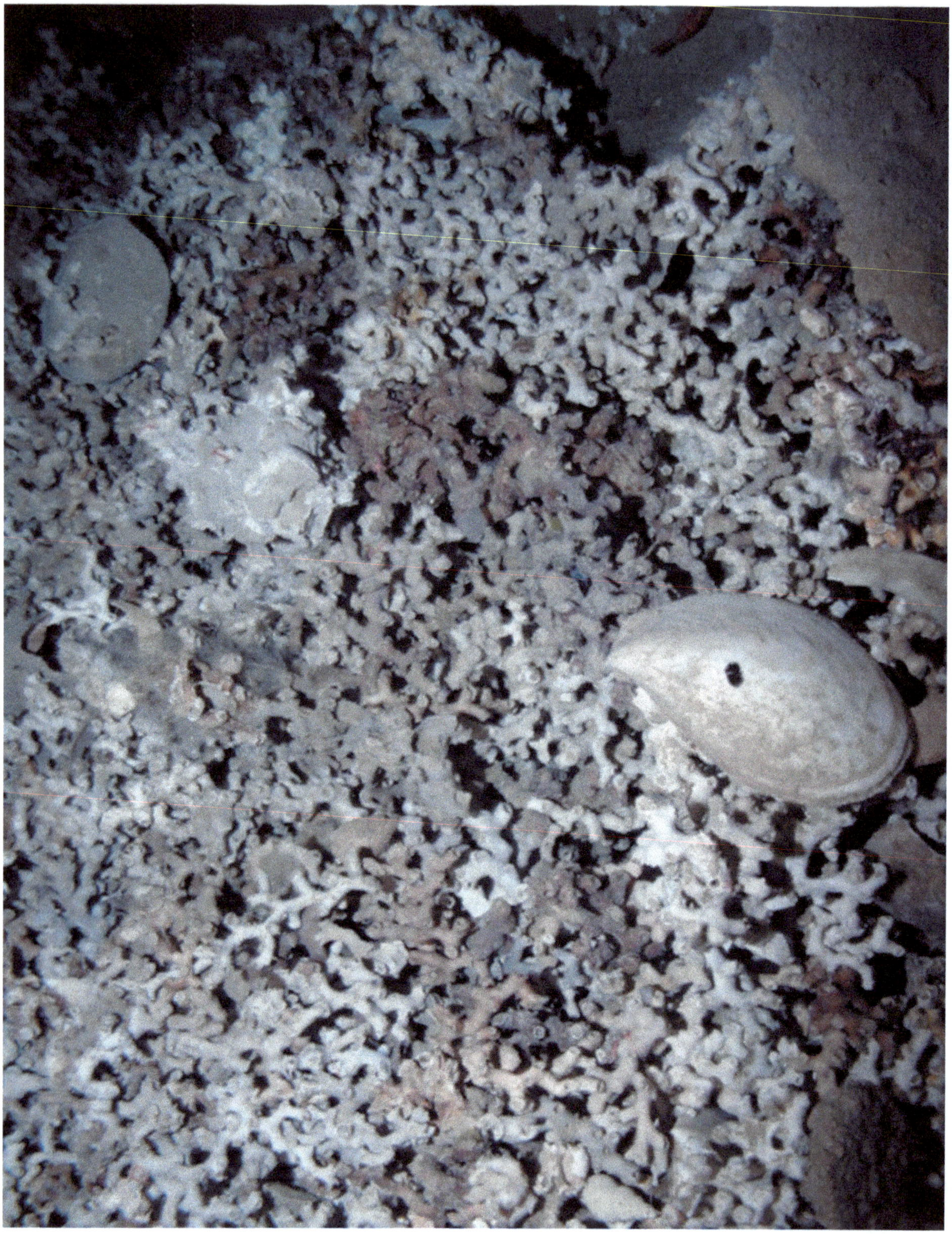

Fig. 9.10 **Biogenic rubble at the foot of an escarpment on the eastern Rockall Bank.** Close-up of coral fragments and bivalve shells at the foot of an escarpment. The field of view is approximately 50 cm across. (Image from the 2009 Offshore Geogenic Reef Mapping Project is provided courtesy of the Department of the Environment, Heritage and Local Government, the Marine Institute and the Geological Survey of Ireland as part of INFOMAR)

Fig. 9.11 **Slabs of rock at the foot of an escarpment on the eastern Rockall Bank.** Slabs of rock that become detached from the cliffs due to ongoing erosion accumulate at the base of the escarpments. The field of view is approximately 2 m across. (Image from the 2009 Offshore Geogenic Reef Mapping Project is provided courtesy of the Department of the Environment, Heritage and Local Government, the Marine Institute and the Geological Survey of Ireland as part of INFOMAR)

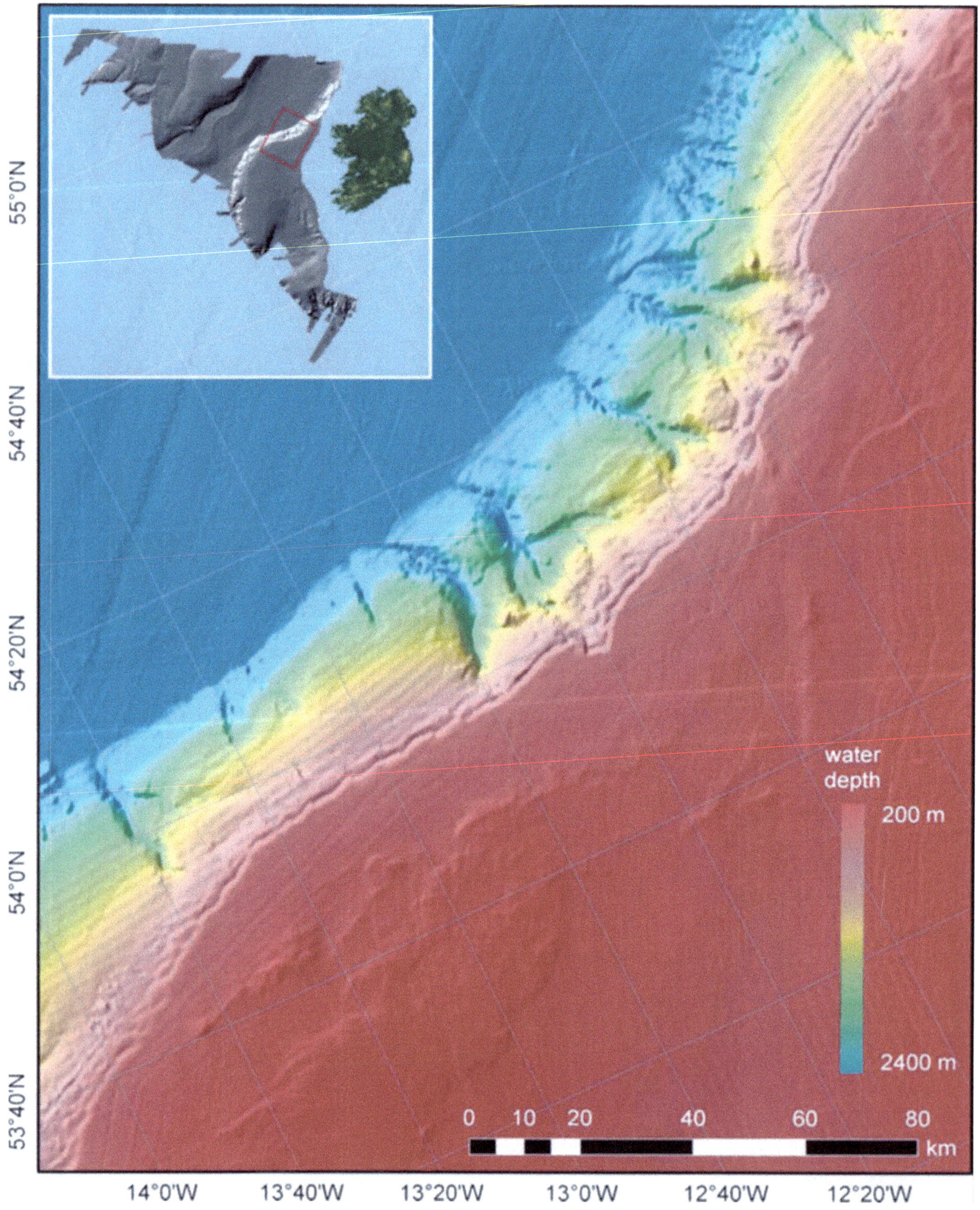

Fig. 9.12 **Escarpments along the north Porcupine Bank.** Escarpments occur in many places along the Porcupine Bank in water depth from ~2,000 to 500 m. Most of the escarpments are, however, confined to the interval between 800 and 500 m water depth. The escarpment in this figure extends for more than 220 km along the northwest edge of Porcupine Bank. At the escarpment, the sea floor drops from 400 to 600 m bsl in the south and from 550 to 750 m bsl in the north

Fig. 9.13 **Fauna on a small escarpment on the northern Porcupine Bank.** Solitary and colonial corals grow upside-down. The image was taken with a forward-looking camera. It clearly illustrates how strong bottom currents erode incompetent sedimentary successions undercutting the more competent units. On top of the competent units, drapes of mobile sand can be found. Sand ripples are indicative of the vigorous bottom current conditions in this area. The field of view is approximately 2 m across. (Image courtesy of the Alfred-Wegener-Institut für Polar- und Meeresforschung and the Institut Français de Recherche pour l'Exploitation de la Mer)

Fig. 9.14 **Fauna on a small escarpment on the northern Porcupine Bank.** In some places solitary and colonial corals occur in high densities. The field of view is approximately 50 cm across. (Image courtesy of the Alfred-Wegener-Institut für Polar- und Meeresforschung and the Institut Français de Recherche pour l'Exploitation de la Mer)

Fig. 9.15 **Fauna on a small escarpment on the northern Porcupine Bank.** A close-up image showing hermit crabs, anemones and sponges living on and in-between the corals. The field of view is approximately 20 cm across. (Image courtesy of the Alfred-Wegener-Institut für Polar- und Meeresforschung and the Institut Français de Recherche pour l'Exploitation de la Mer)

Fig. 9.16 **Fauna on a small escarpment on the northern Porcupine Bank.** Increased numbers of fish occur along the escarpments. The field of view is approximately 2 m across. (Image courtesy of the Alfred-Wegener-Institut für Polar- und Meeresforschung and the Institut Français de Recherche pour l'Exploitation de la Mer)

Bibliography

Dorschel, B., Wheeler, A. J., Huvenne, V. A. I., & de Haas, H. (2009). Cold-water coral mounds in an erosive environmental setting: TOBI side-scan sonar data and ROV video footage from the northwest Porcupine Bank, NE Atlantic. *Marine Geology, 264*, 218–229.

Elliott, G. M., Shannon, P. M., Haughton, P. D. W., & Øvrebø, L. K. (2010). The Rockall Bank mass flow: Collapse of a moated contourite drift onlapping the eastern flank of Rockall Bank, west of Ireland. *Marine and Petroleum Geology, 27*(1), 92–107.

Weaver, P. P. E., Wynn, R. B., Kenyon, N. H., & Evans, J. (2000). Continential margin sedimentation, with special reference to the north-east Atlantic margin. *Sedimentology, 47*(s1), 239–256.

Wheeler, A. J., Beck, T., Thiede, J., Klages, M., Grehan, A., Monteys, F. X., et al. (2005). Deep-water coral mounds on the Porcupine Bank, Irish margin: preliminary results from Polarstern ARK-XIX/3a ROV cruise. In A. Freiwald & J. M. Roberts (Eds.), *Cold-water corals and ecosystems* (pp. 323–333). Berlin: Springer.

Mound Features and Coral Carbonate Mounds

10

B. Dorschel, A.J. Wheeler, X. Monteys, and K. Verbruggen

Mound features, in the broadest sense, describe morphological seabed features that measure hundreds of metres to several kilometres across and elevate tens to hundreds of metres above the surrounding seabed (Fig. 10.1). They are distinct from seamounts (Chapter 8) which are much larger. The shapes of these features can vary from simple conical, to elongated and even complex multi-summited structures. The term mound feature in the context of this atlas describes features of comparable morphology without differentiating on the basis of their origins (as groundtruthing data is often scarce). Therefore, the mound features mapped here include those of geologic origin (such as volcanic structures and rock outcrops) as well as biogenic structures such as cold-water coral carbonate mounds. Even though the mound features have a variety of origins, they all elevate from the seabed and protrude into the water column forming obstacles to bottom currents. Thus, they are often subject to erosion, providing hardgrounds and environmental conditions for organisms to settle and dwell.

Mound features of volcanic origin, such as the cluster of mounds at the foot of the northwest corner of the Porcupine Bank, can be described as mini-seamounts and represent the remnants of submarine volcanoes (Fig.10.15). Located in an area of intensified bottom currents, these mound features remain exposed and have not been covered by sediments. Rock outcrops are also restricted to areas of intense bottom currents where erosion exposes these features and prevents their burial. In addition to these geological features, the Irish seabed also hosts numerous biogenic mound features, the so-called cold-water coral carbonate mounds. These mounds are entirely made up of loose frameworks of coral skeletons embedded in a matrix of background (hemipelagic) sediments. They are simply the result of persistent occurrences of cold-water corals that trap sediment and build on the remains of former corals. In some places, these mounds can rise more than 300 m above the surrounding seafloor i.e. more than twice the size of the Great Pyramid of Giza (140 m high).

Coral carbonate mounds have been found in many places in the world oceans. To date they have been reported from both sides of the Atlantic (from northern Norway

B. Dorschel (✉) and A.J. Wheeler
Department of Geology & Environmental Research Institute, University College Cork, Cork, Ireland
e-mails: b.dorschel@ucc.ie; a.wheeler@ucc.ie

X. Monteys and K. Verbruggen
Geological Survey of Ireland, Beggars Bush, Haddington Road, Dublin 4, Ireland
e-mails: Xavier.Monteys@gsi.ie; Koen.Verbruggen@gsi.ie

B. Dorschel et al., *Atlas of the Deep-Water Seabed: Ireland*,
DOI 10.1007/978-90-481-9376-9_10, © Springer Science+Business Media B.V. 2010

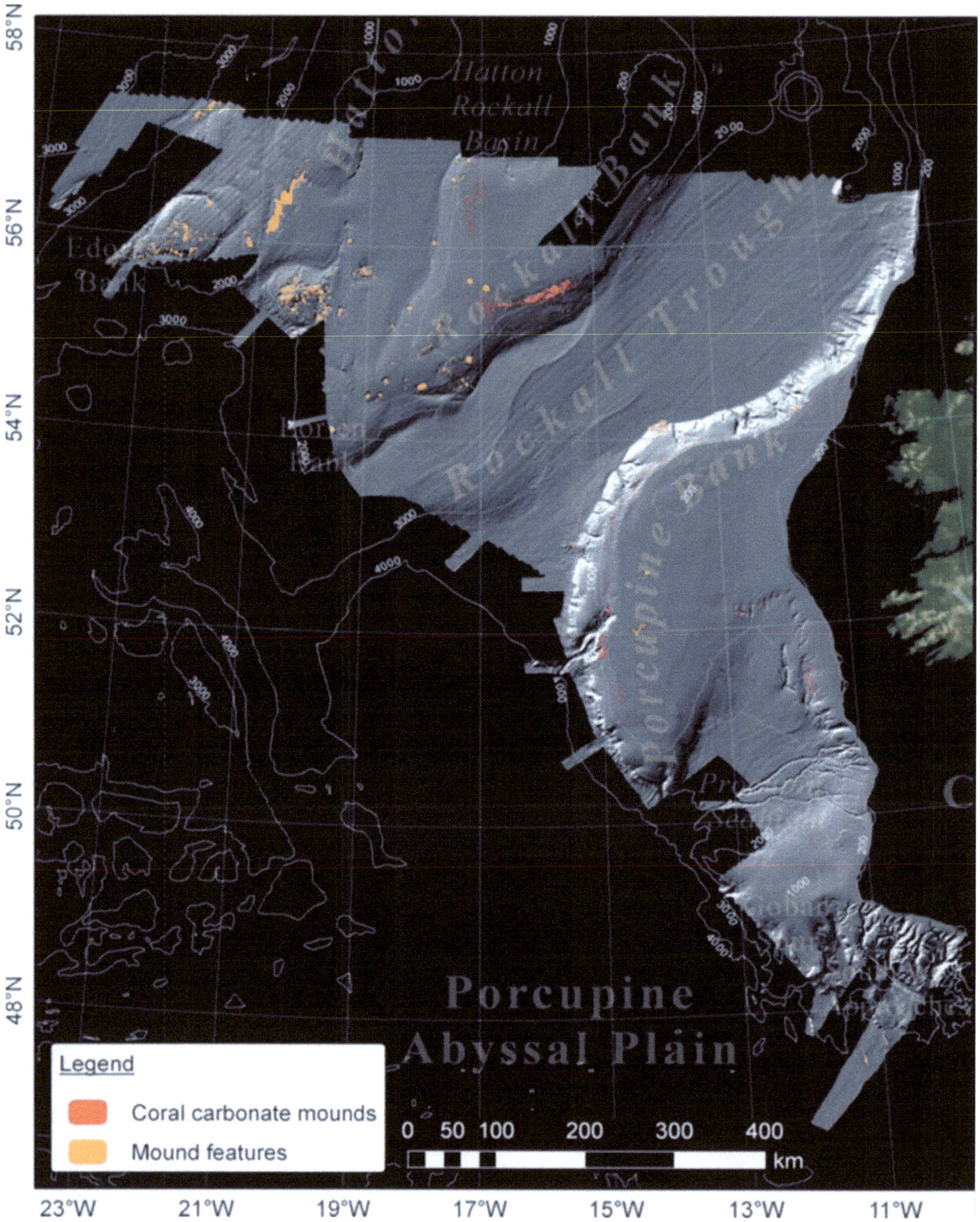

Fig. 10.1 **Mound features & coral carbonate mounds.** Mound features are widely distributed on the Irish seabed and occur from the Porcupine Seabight to the Hatton Bank. These features can be of geological origin (e.g. volcanic structures, rock outcrops) or biological origin (coral carbonate mounds). Examples of mounds of volcanic origin can be found at the northwest corner of the Porcupine Bank. Coral carbonate mounds mainly occur between 500 and 1,500 m water depth along Rockall Bank, the western side of Porcupine Bank and the eastern Porcupine Seabight. In addition to these mound features, numerous mound features of unknown origin have been recorded. In order to identify the origin of a mound feature, video and seismic information are necessary

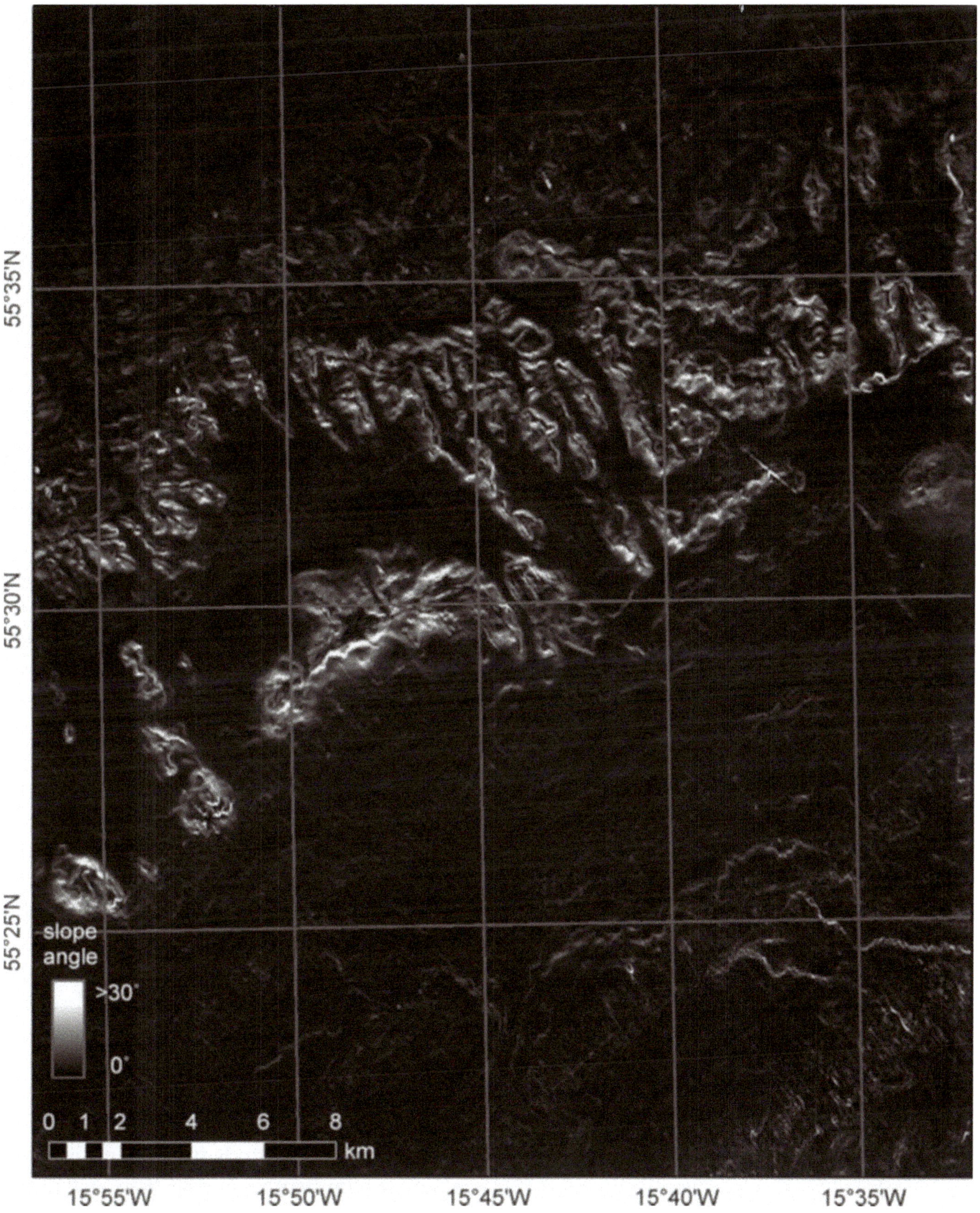

Fig. 10.2 **Mound detection.** Mounds were detected based on inclination images. The slopes of the mounds are significantly steeper than the normal seabed. Therefore, the slopes of mounds show up well on slope inclination raster maps. The image shows mounds on the eastern slope of the Rockall Bank (see also Fig. 10.3)

to Mauritania and from Brazil to Canada) and from New Zealand waters. The Irish seabed, however, hosts the most abundant and the largest cold-water coral mounds known so far.

Perhaps surprising, corals are quite common on the Irish seabed. However, they are different from tropical corals in many ways. One of the most important differences is that these coral do not have photosynthesising algal symbionts and are therefore not restricted to the photic zone but can live down to several thousand metres water depth. They are furthermore not restricted to warm waters. Irish cold-water corals, as their name suggests, prefer water temperatures between 4°C and 12°C. Like tropical corals, some cold-water corals have skeletons of calcium carbonate and can form reefs.

On the Irish seabed, the two reef-forming cold-water coral species *Lophelia pertusa* and *Madrepora oculata* are widely distributed, dwelling on available hard substrate. In areas that suit them, cold-water corals form thickets and reefs. Within the thickets and reefs, sediment accumulation is increased due to the baffling capacity of the corals and due to additional carbonate added to the sediments by the corals. If cold-water coral reef growth repeatedly occurs in a particular location over geological time periods, it locally increases the sediment accumulation outpacing sedimentation in off-mound settings. The result is an elevated morphological feature: a mound.

Coral carbonate mounds have been a focus of modern scientific research since the early 1990s and a large amount of information on coral carbonate mounds has been gathered since. In recent years, several research cruises have targeted coral carbonate mounds mapping, videoing and sampling them. As a result, groundtruthing data are available for many mound features in the Porcupine Seabight and along the Porcupine and Rockall Bank. Based on these groundtruthing data, coral carbonate mounds have been singled out and are present as an individual sub-set of the overall mound features (Fig. 10.1).

10.1 Identification

As a first step, the summits of mound features were marked and the coordinates and water depths of these marker points were exported as latitude, longitude and depth value tables. The detection of the mound bases turned out to be somewhat more difficult than that of the mound summits. Mounds are not always located on flat and even terrain. They often occur on steep slopes, at canyon heads, along canyon walls or on top of escarpments and are superimposed on larger morphological features such as seamounts, banks and ridges (Fig. 10.2). Therefore, mound bases were defined by a critical slope angle of 6°. The 6° slope contour in combination with bathymetric data allowed for the manual detection of the bases of all mound features. The threshold value of 6° was chosen as a compromise between the increasing influence of artefacts at smaller slope angles and the under-detection of mounds at larger slope angles. Out of all the mounds identified, cold-water coral carbonate mounds were identified only where groundtruthing data was available. Thus, the coral carbonate mound dataset is a sub-set of the mound dataset.

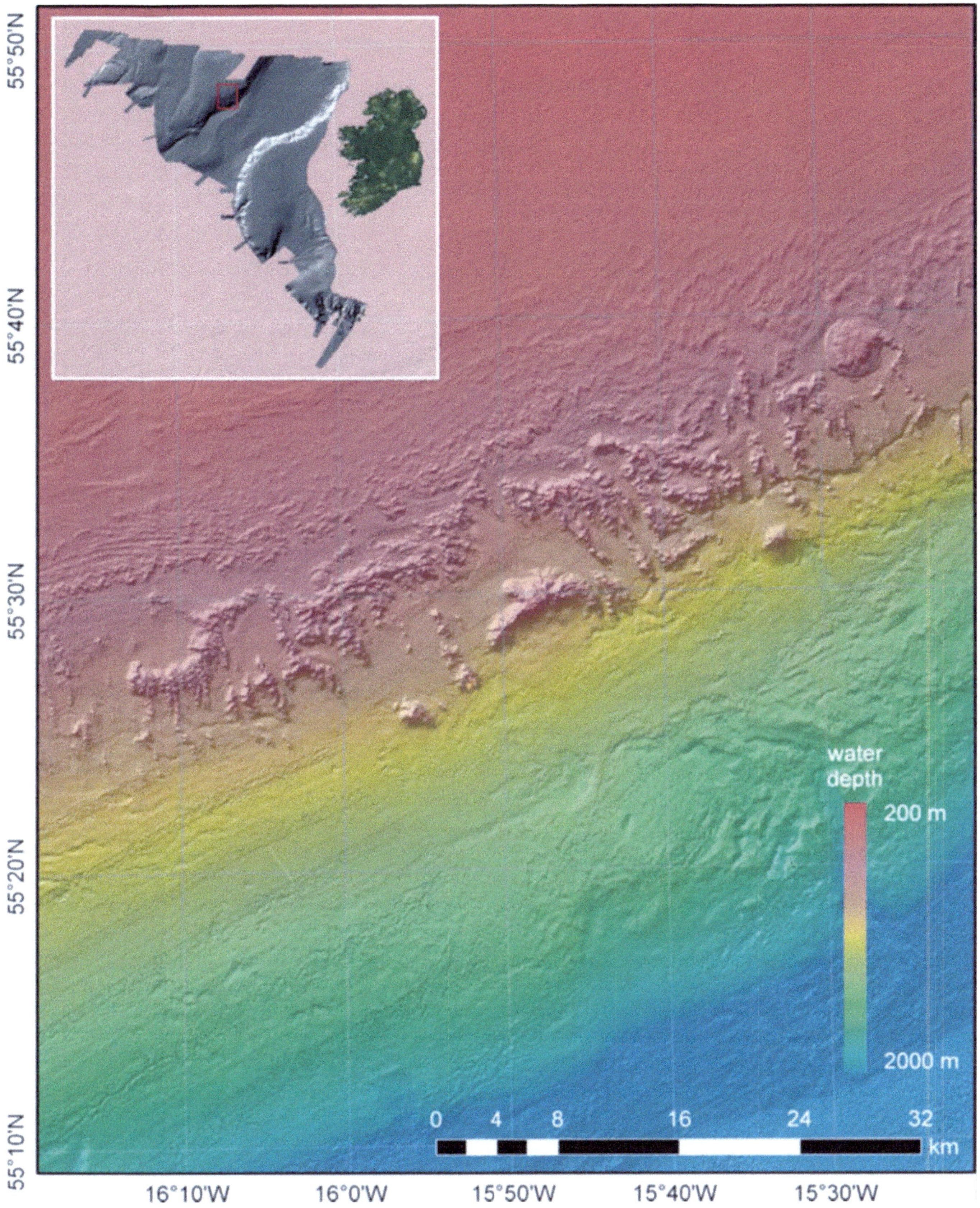

Fig. 10.3 Coral carbonate mounds on the eastern slope of the Rockall Bank. The Logachev mound province, located on the eastern slopes of the Rockall Bank, contains what are probably the largest coral carbonate mounds in the world. The tallest of these mounds elevate for several hundreds of metres above the surrounding seafloor

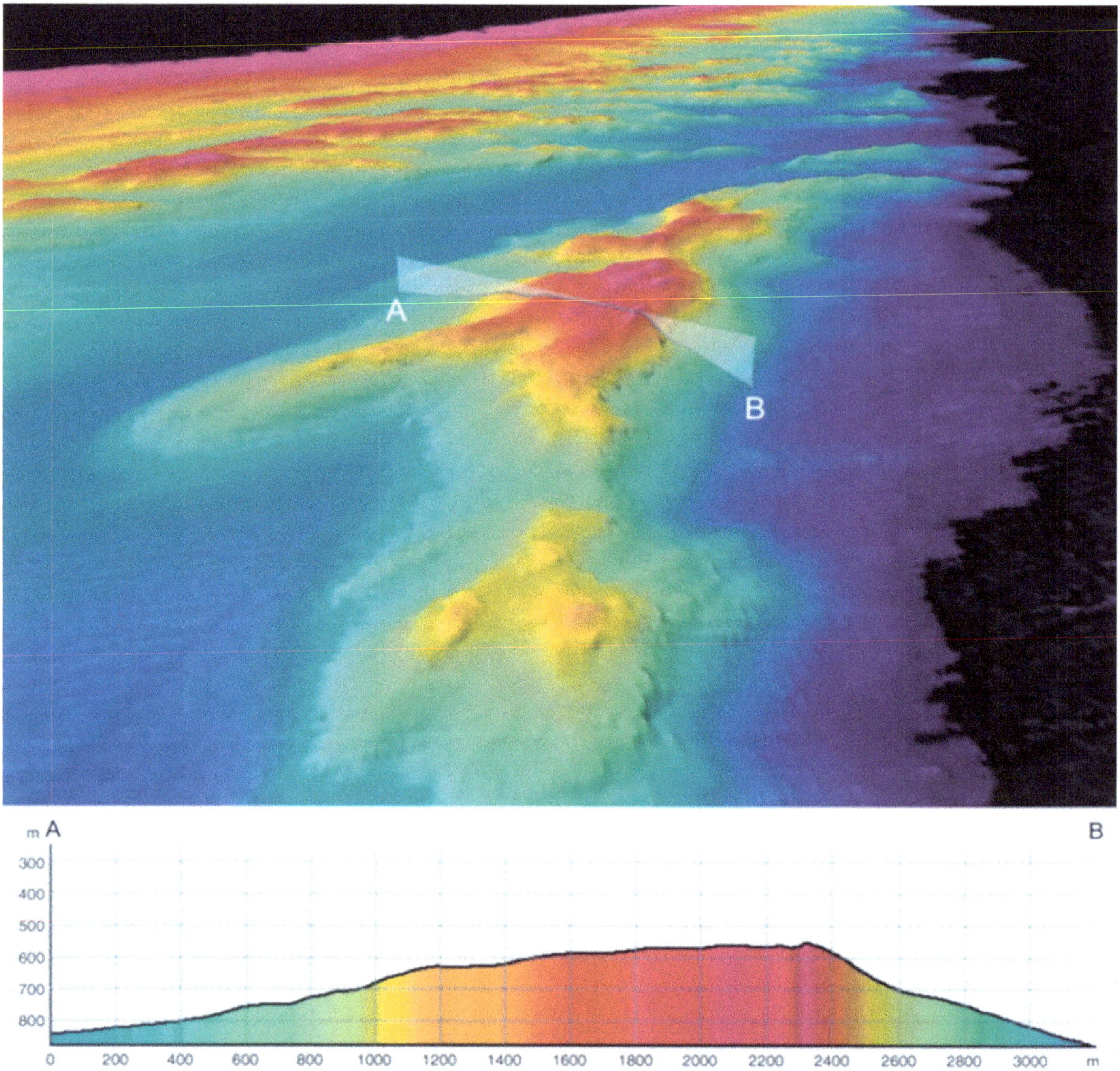

Fig. 10.4 Bathymetric profile across a coral carbonate mound on the upper eastern Rockall Bank. The three-dimensional image shows part of the Logachev mound province on the eastern slopes of Rockall Bank. The line A-B marks the track of the bathymetric profile across one individual mound shown in the image below. 3D image and profile are not vertical exaggerated and give therefore an impression of the true steepness of the seabed in this area

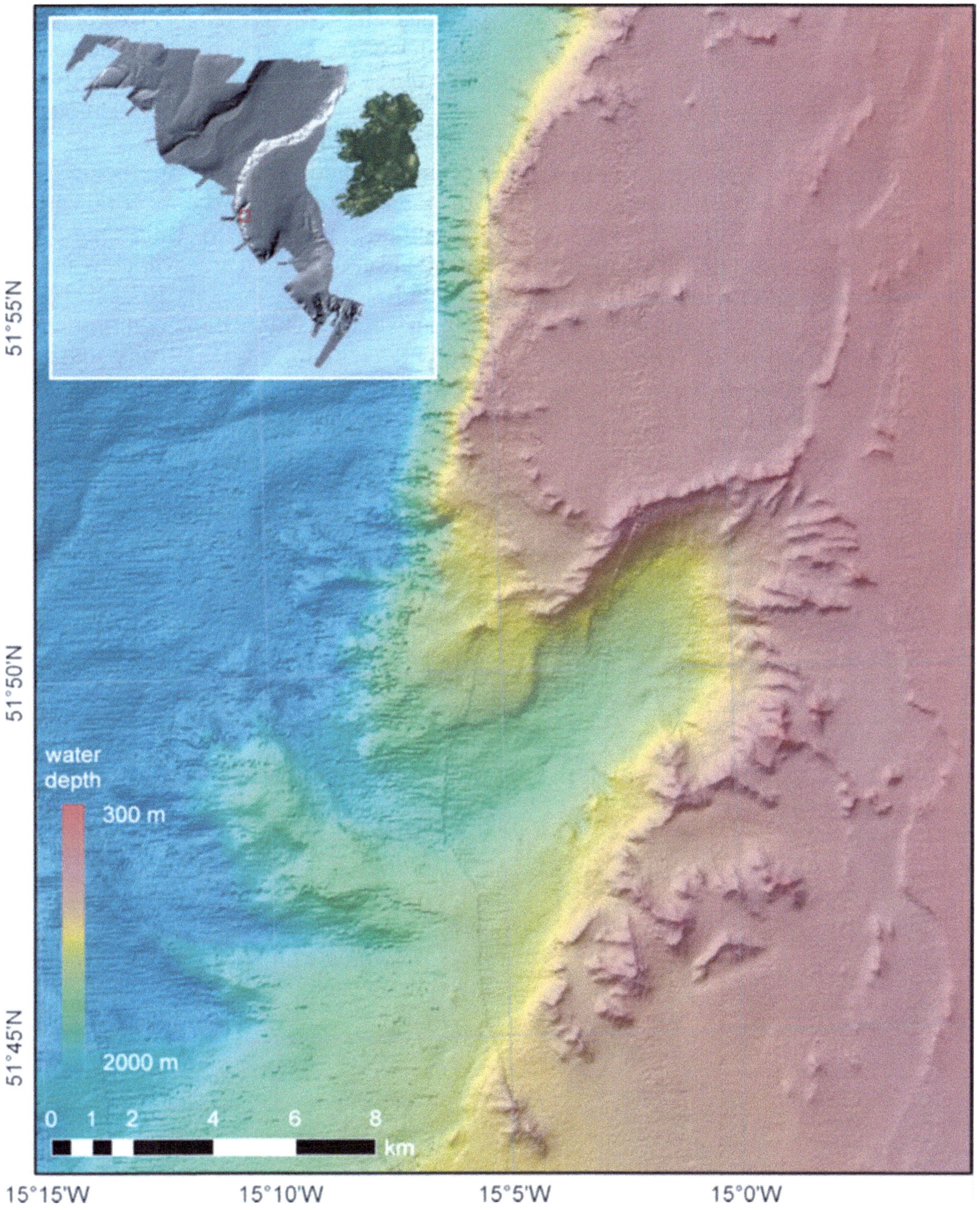

Fig. 10.5 Coral carbonate mounds along a canyon on the western slope of the Porcupine Bank. Coral carbonate mounds often occur on or at canyon heads and the top of walls and escarpments. These areas are favourable environments for cold-water corals to thrive

10.2 Distribution

Mound features occur in many places on the Irish seabed from west of the SW Approaches to the slopes of Hatton Bank. Apart from some individual mound features, most occur in distinct mound clusters and provinces (Fig. 10.1).

The Porcupine Seabight for example hosts three well-studied provinces of coral carbonate mounds (Fig. 10.6). Further coral carbonate mound provinces are known from the Porcupine and Rockall Banks (Fig. 10.3 & 10.5). Coral carbonate mounds occur almost exclusively between 500 and 1,500 m water depth. As a conservative estimate, almost two thirds of the 1,549 Irish mound features can be assigned to coral carbonate mounds.

Other mound features such as the mini-seamounts and mound features below 1,500 m water depth are clearly not coral carbonate mounds (Fig. 10.15). Unfortunately no groundtruthing data are available for the mound features in the Rockall-Hatton Basin, on the Hatton Bank and on the seamounts (Fig. 10.16). It is likely that these features include coral carbonate mounds as well as mounds of different origin.

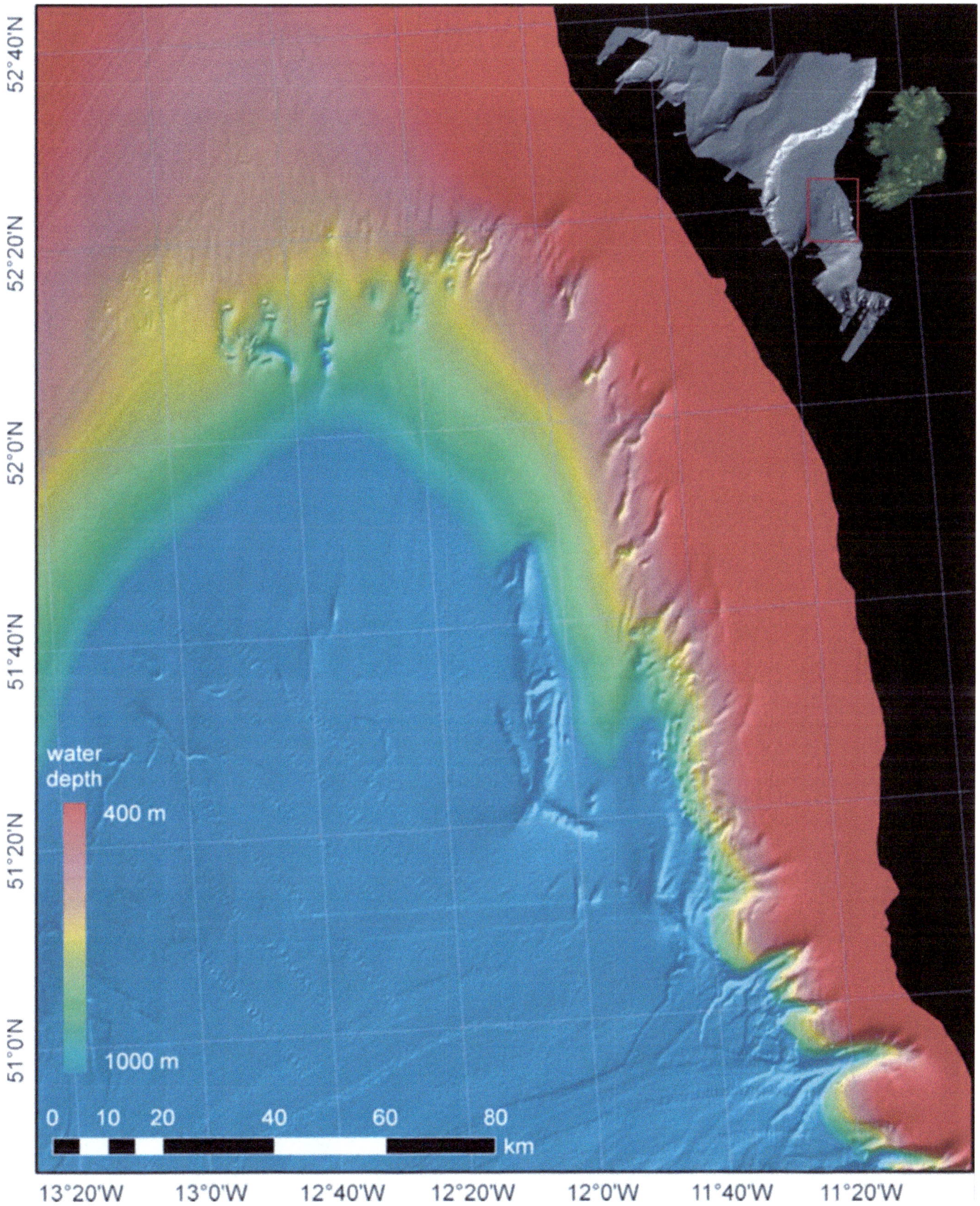

Fig. 10.6 **Provinces of coral carbonate mounds in the Porcupine Seabight.** The Porcupine Seabight contains well investigated coral carbonate mound provinces: the Hovland mound province in the northern Porcupine Seabight (*top left corner*) and the Belgica mound province in the eastern Porcupine Seabight (*bottom right corner*). In addition to these two mound provinces, the Porcupine Seabight also contains a province of buried mounds located just to the north of the Hovland mound province (the Magellan mounds)

Fig. 10.7 **A coral thicket on a coral carbonate mound.** Some of the coral carbonate mounds are overgrown by thickets of cold-water corals. This image shows a colony of *Lophelia pertusa*, one of the main mound-building corals on the Irish seabed.The field of view is approximately 2 m across (Image courtesy of MARUM – Center for Marine Environmental Sciences, University of Bremen)

Fig. 10.8 **A coral thicket on a coral carbonate mound.** Large sponges also occur on coral carbonate mounds. These mounds are furthermore potential nursing areas for many fish species. The field of view is approximately 2 m across (Image courtesy of MARUM – Center for Marine Environmental Sciences, University of Bremen)

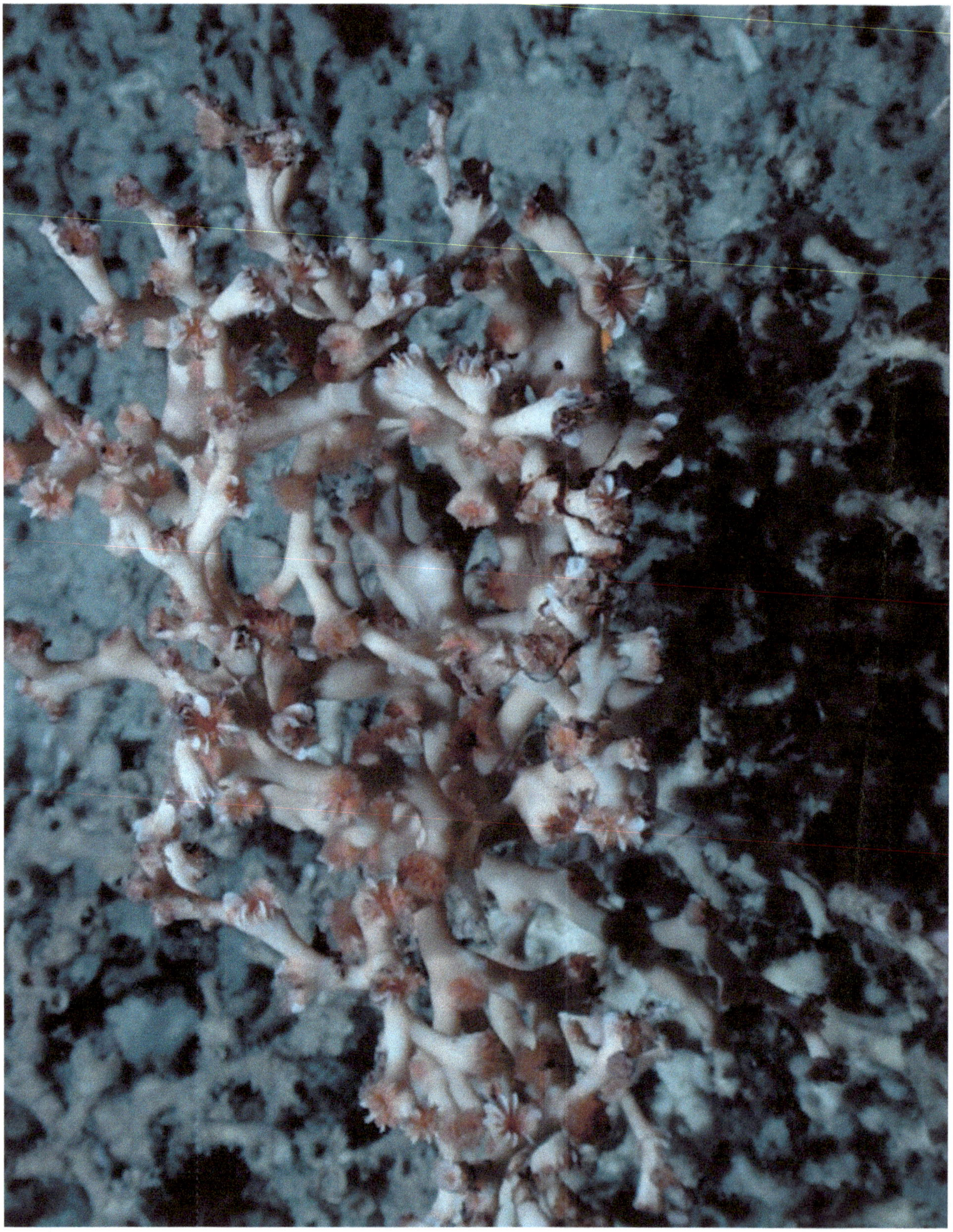

Fig. 10.9 A close look at a cold-water coral. The actual living coral organisms are the orange areas in the centres of the thecae at the ends of the coral branches. The whitish parts of the corals are their aragonitic (calcium carbonate) skeletons. The field of view is approximately 30 cm across (Image courtesy of MARUM – Center for Marine Environmental Sciences, University of Bremen)

Fig. 10.10 Mound initiation. A setting like this may represent the very early start of a coral carbonate mound. Available hard substrate like this rock becomes quickly colonised if the environmental conditions are favourable. The field of view is approximately 2 m across (Image courtesy of the Alfred-Wegener-Institut für Polar- und Meeresforschung and the Institut Français de Recherche pour l'Exploitation de la Mer)

Fig. 10.11 **An eroded coral carbonate mound.** When the corals die and the mound starts eroding, consolidated and partly cemented sediments become exposed in which exhumed fossil coral fragments can still be distinguished. The old coral skeletons provide substrate for other organisms. The field of view is approximately 30 cm across (Image courtesy of MARUM – Center for Marine Environmental Sciences, University of Bremen)

Fig. 10.12 Fauna associated with coral carbonate mounds. Hermit crabs and squids found on a soft coral. The field of view is approximately 50 cm across (Image courtesy of MARUM – Center for Marine Environmental Sciences, University of Bremen)

Fig. 10.13 **Fauna associated with coral carbonate mounds.** Anemones and hermit crabs, hydrozoans and soft corals. The field of view is approximately 50 cm across (Image courtesy of MARUM – Center for Marine Environmental Sciences, University of Bremen)

Fig. 10.14 **Fauna associated with coral carbonate mounds.** A black coral (coloured bright orange) with a galatheid crab. The field of view is approximately 2 m across (Image courtesy of MARUM – Center for Marine Environmental Sciences, University of Bremen)

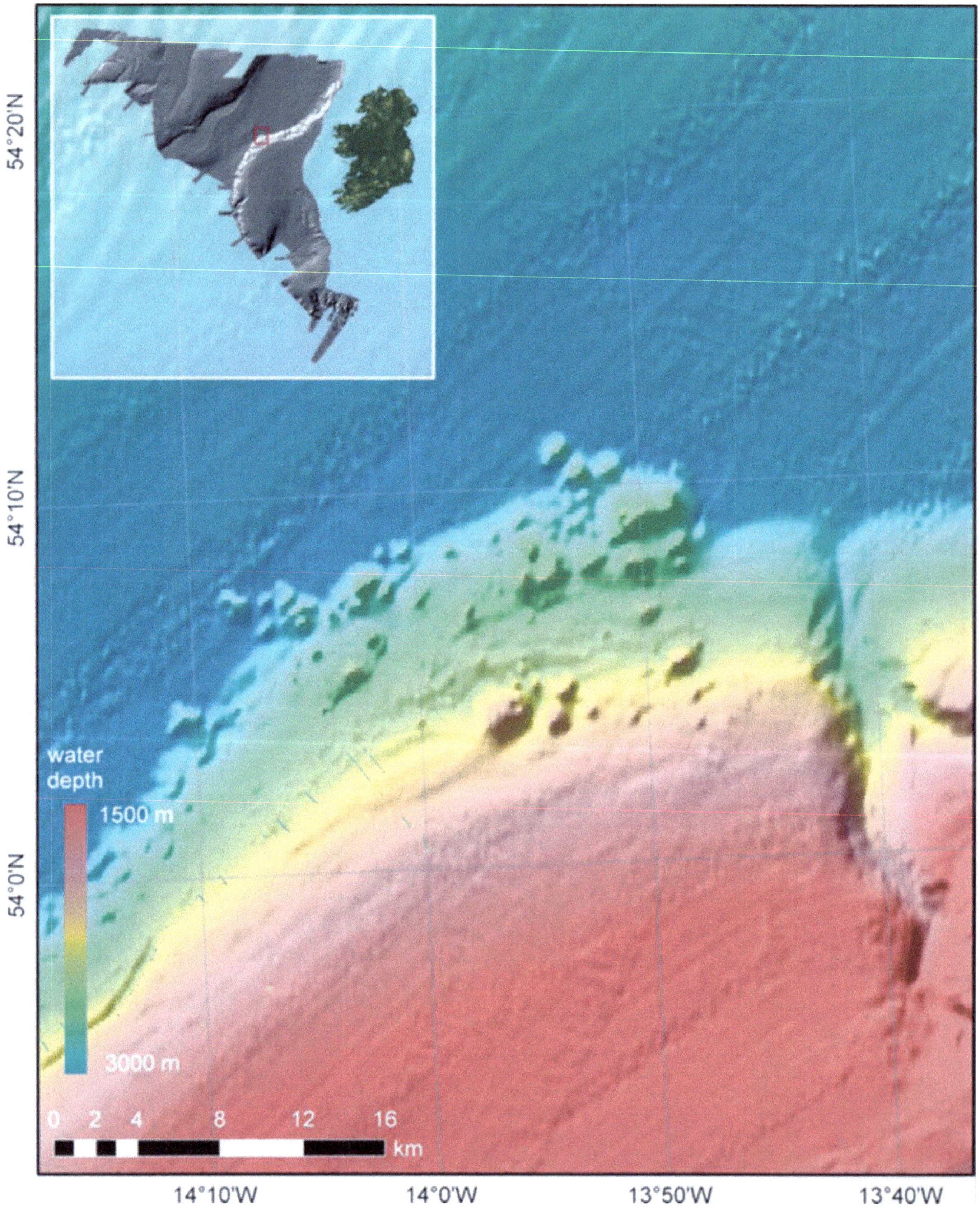

Fig. 10.15 **Mounds of volcanic origin, mini-seamounts, at the north-west corner of the Porcupine Bank.** These small structures are of volcanic origin thus representing small seamounts. Located between 2,000 and 3,000 m bsl, these features also highlight the limits of the bathymetric dataset. The small kilometre-sized features are located in deep water (deeper than 2,000 m bsl). The decreasing data density with increasing water depth also decreases the resolution of the DEM and hence the degraded quality of the image

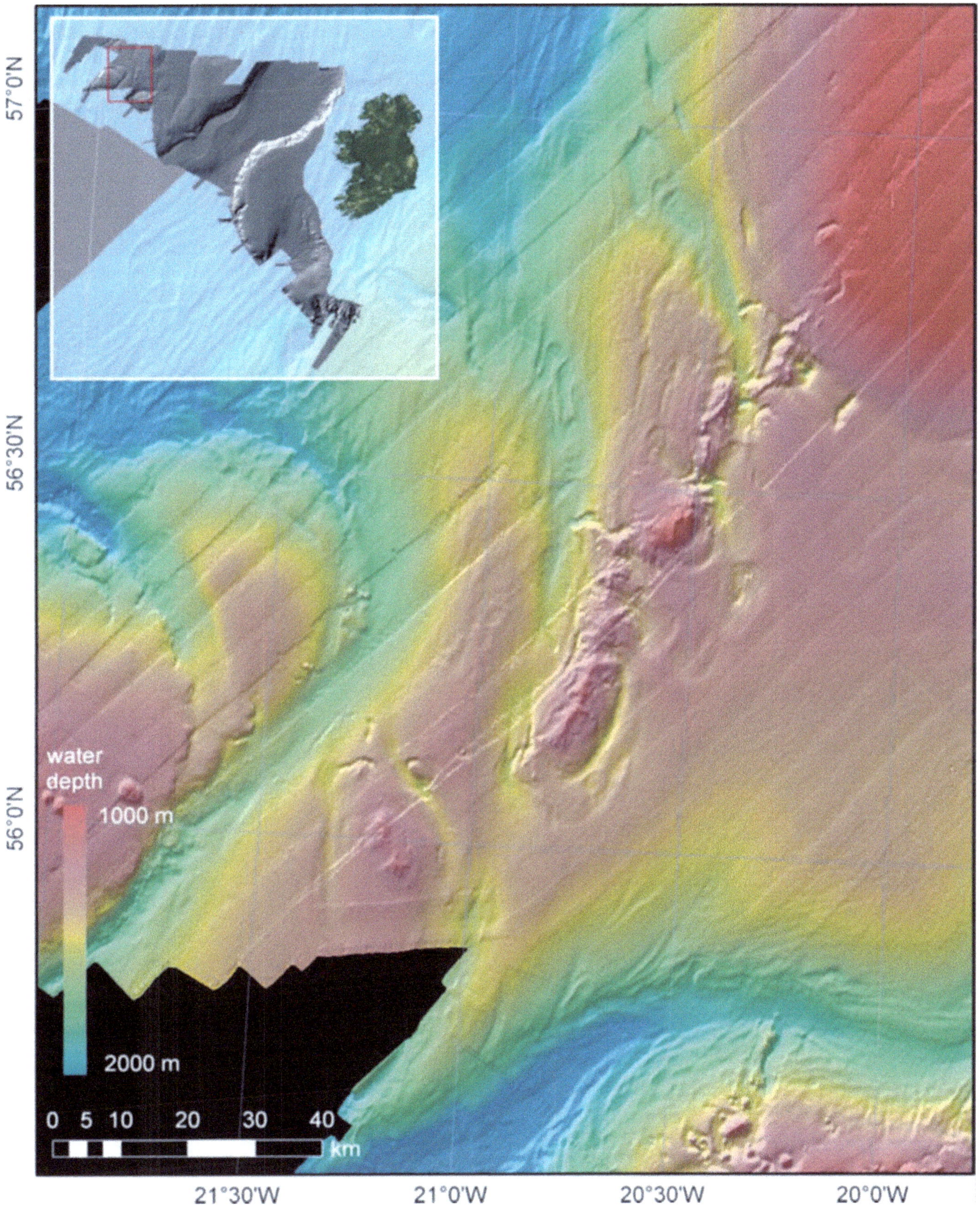

Fig. 10.16 Mounds of unknown origin south of the Hatton Bank. These mounds are located in water depth of approximately 1,500 m bsl. Due to the lack of groundtruthing information it is unclear what type of mound feature is represented here

Bibliography

de Haas, H., Mienis, F., Frank, N., Richter, T. O., Steinacher, R., de Stigter, H., et al. (2009). Morphology and sedimentology of (clustered) cold water coral mounds at the south Rockall Trough margins, NE Atlantic Ocean. *Facies, 55*, 1–26.

De Mol, B., Van Rensbergen, P., Pillen, S., Van Herreweghe, K., Van Rooij, D., McDonnell, A., et al. (2002). Large deep-water coral banks in the Porcupine Basin, southwest of Ireland. *Marine Geology, 188*(1–2), 193–231.

Dorschel, B., Wheeler, A. J., Huvenne, V. A. I., & de Haas, H. (2009). Cold-water coral mounds in an erosive environmental setting: TOBI side-scan sonar data and ROV video footage from the northwest Porcupine Bank, NE Atlantic. *Marine Geology, 264*, 218–229.

Freiwald, A., & Roberts, J. M. (2005). *Cold-water corals and ecosystems*. Berlin/Heidelberg/New York: Springer. 1243 pp.

Kenyon, N. H., Akhmetzhanov, A. M., Wheeler, A. J., van Weering, T. C. E., de Haas, H., & Ivanov, M. K. (2003). Giant carbonate mud mounds in the southern Rockall Trough. *Marine Geology, 195*, 5–30.

Mienis, F., van Weering, T. C. E., de Haas, H., de Stigter, H. C., Huvenne, V. A. I., & Wheeler, A. J. (2006). Carbonate mound development at the SW Rockall Trough margin based on high resolution TOBI and seismic recording. *Marine Geology, 233*(1–4), 1–19.

Roberts, J. M., Wheeler, A. J., Cairns, S., & Freiwald, A. (2009). *Cold-water corals: The biology and geology of deep-sea coral habitats* (p. 336). Cambridge: Cambridge University Press.

Roberts, J. M., Wheeler, A. J., & Freiwald, A. (2006). Reefs of the deep: The biology and geology of cold-water coral ecosystems. *Science, 312*, 543–547.

Wheeler, A. J., Beyer, A., Freiwald, A., de Haas, H., Huvenne, V. A. I., Kozachenko, M., et al. (2007). Morphology and environment of deep-water coral mounds on the NW European Margin. *International Journal of Earth Sciences, 96*, 37–56.

van Weering, T. C. E., de Haas, H., de Stigter, H. C., Lykke-Andersen, H., & Kouvaev, I. (2003). Structure and development of giant carbonate mounds at the SW and SE Rockall Trough margins, NE Atlantic Ocean. *Marine Geology, 198*, 67–81.

Iceberg Ploughmarks

11

B. Dorschel, A.J. Wheeler, X. Monteys, and K. Verbruggen

Iceberg ploughmarks are caused by icebergs that ground on the seafloor. Today, icebergs are very rare in Irish waters. This, however, was not always the case. During past glacials (e.g. 20,000 years ago) numerous icebergs reached Ireland's shores leaving their marks on the seabed (Fig. 11.1). Typically, icebergs are moved either by ocean currents or winds. Once grounded, the bottom of the iceberg scraps the seabed as the iceberg continues to move. High tides can lift the icebergs temporarily clear of the seabed but often dropping them again before they have had a chance to move very far, in many cases they are merely moved upslope into even shallower waters. Melting of icebergs reduces their draft thus allowing the iceberg to either drift away or get dragged even further up the slope. The result is that icebergs leave marks in the soft sediments of the seafloor that usually follow gentle zigzag lines climbing up the slopes (Fig. 11.2–11.4).

When the grounded icebergs move, they plough through the sediments leaving a ridge of sediments to the left and to the right of a central furrow. Iceberg ploughmarks are therefore elongated depressions bordered by sediments levees on either side (Fig. 11.3). When melting, icebergs also release some of the sediment load (so-called ice-rafted debris) that was frozen in by the glacial ice eroding the landmasses. Therefore, large erratic boulders are common in the areas of iceberg ploughmarks.

Based on the water depth where iceberg ploughmarks occur, it is possible to calculate the draft of the icebergs and hence their overall sizes. To generate a conservative estimate, we assume that the icebergs were grounding when sea level was the lowest during the glacial maxima (about 120–130 m lower than today). The deepest iceberg ploughmark is found at a recent water depth of 580 m bsl at the western slope of the Rockall Bank. The iceberg that caused this mark had therefore a minimum draft of 450 m. With approximately 7/8th of an iceberg below the water and only 1/8th above, the total size of this iceberg must have been more than 514 m tall – that's more than one and a half times as tall as the Eiffel Tower.

As small features in soft sediments, iceberg ploughmarks have only a limited preservation potential. In areas of enhanced bottom currents they disappear quickly as

B. Dorschel (✉) and A.J. Wheeler
Department of Geology & Environmental Research Institute, University College Cork, Cork, Ireland
e-mails: b.dorschel@ucc.ie; a.wheeler@ucc.ie

X. Monteys and K. Verbruggen
Geological Survey of Ireland, Beggars Bush, Haddington Road, Dublin 4, Ireland
e-mails: Xavier.Monteys@gsi.ie; Koen.Verbruggen@gsi.ie

B. Dorschel et al., *Atlas of the Deep-Water Seabed: Ireland*,
DOI 10.1007/978-90-481-9376-9_11, © Springer Science+Business Media B.V. 2010

Fig. 11.1 Iceberg ploughmarks. Iceberg ploughmarks occur on all the major submarine banks and on the continental slope. Evidence for icebergs is found in all areas shallower than approximately 500 m water depth

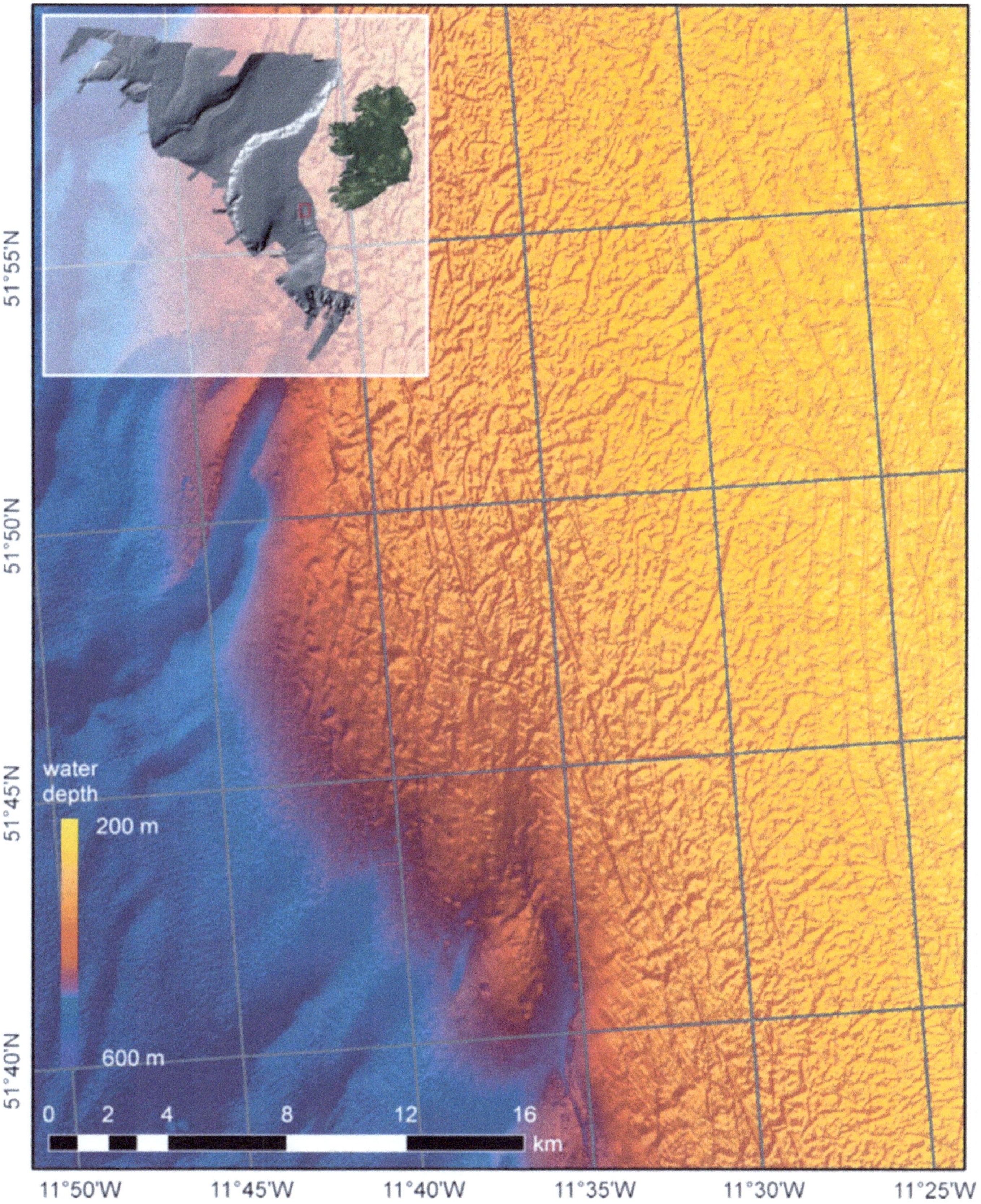

Fig. 11.2 Iceberg ploughmarks in the Porcupine Seabight. Iceberg ploughmarks on the eastern slope of the Porcupine Seabight. The image has no vertical exaggeration. For stronger contrasts of small seabed features, the hillshade effect is 100 times amplified. The parallel lineations are artefacts in the outer beams of the multibeam swath

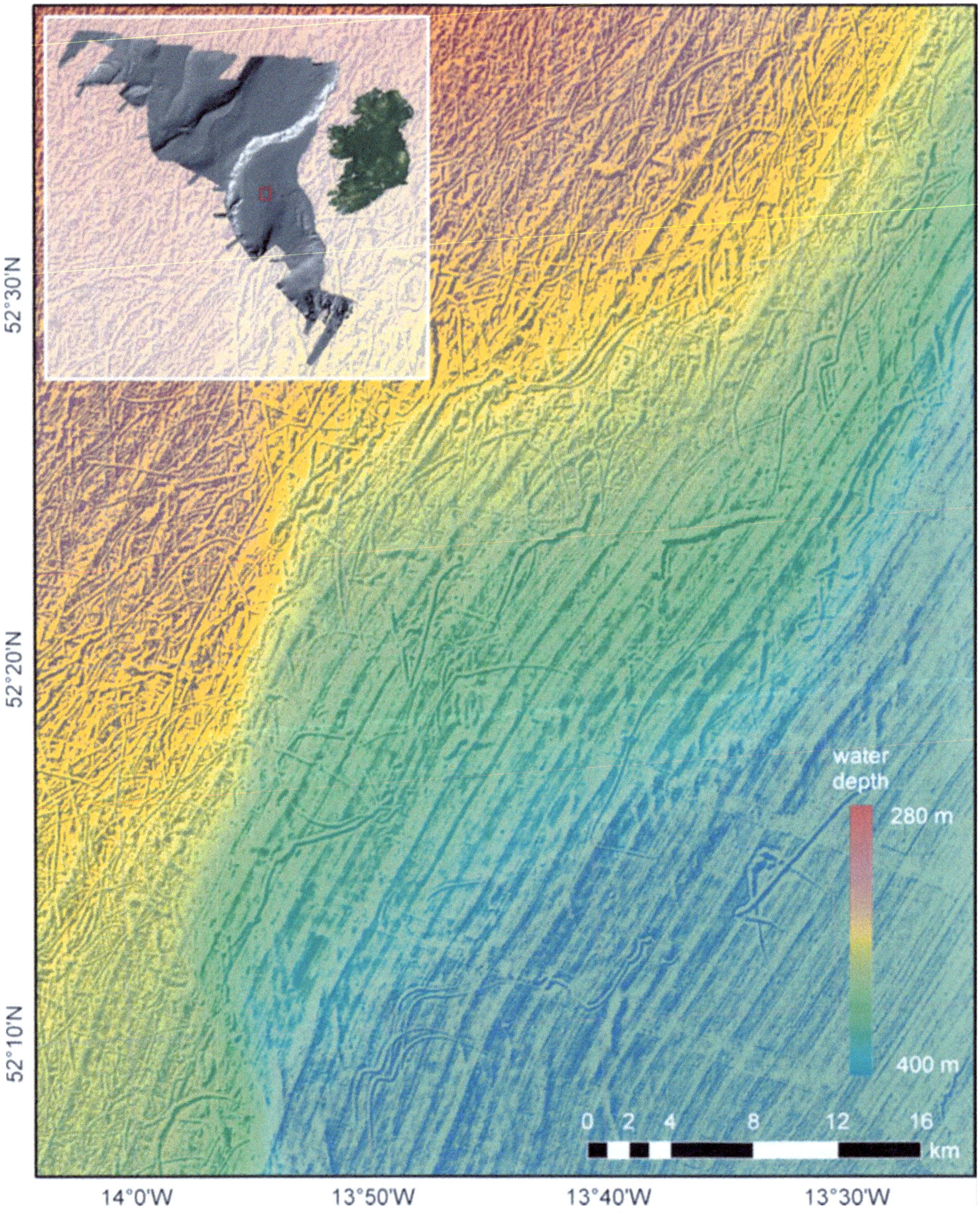

Fig. 11.3 Iceberg ploughmarks in the Porcupine Seabight. Iceberg ploughmarks are common in the Porcupine Seabight, with individual ploughmarks traced for tens of kilometres

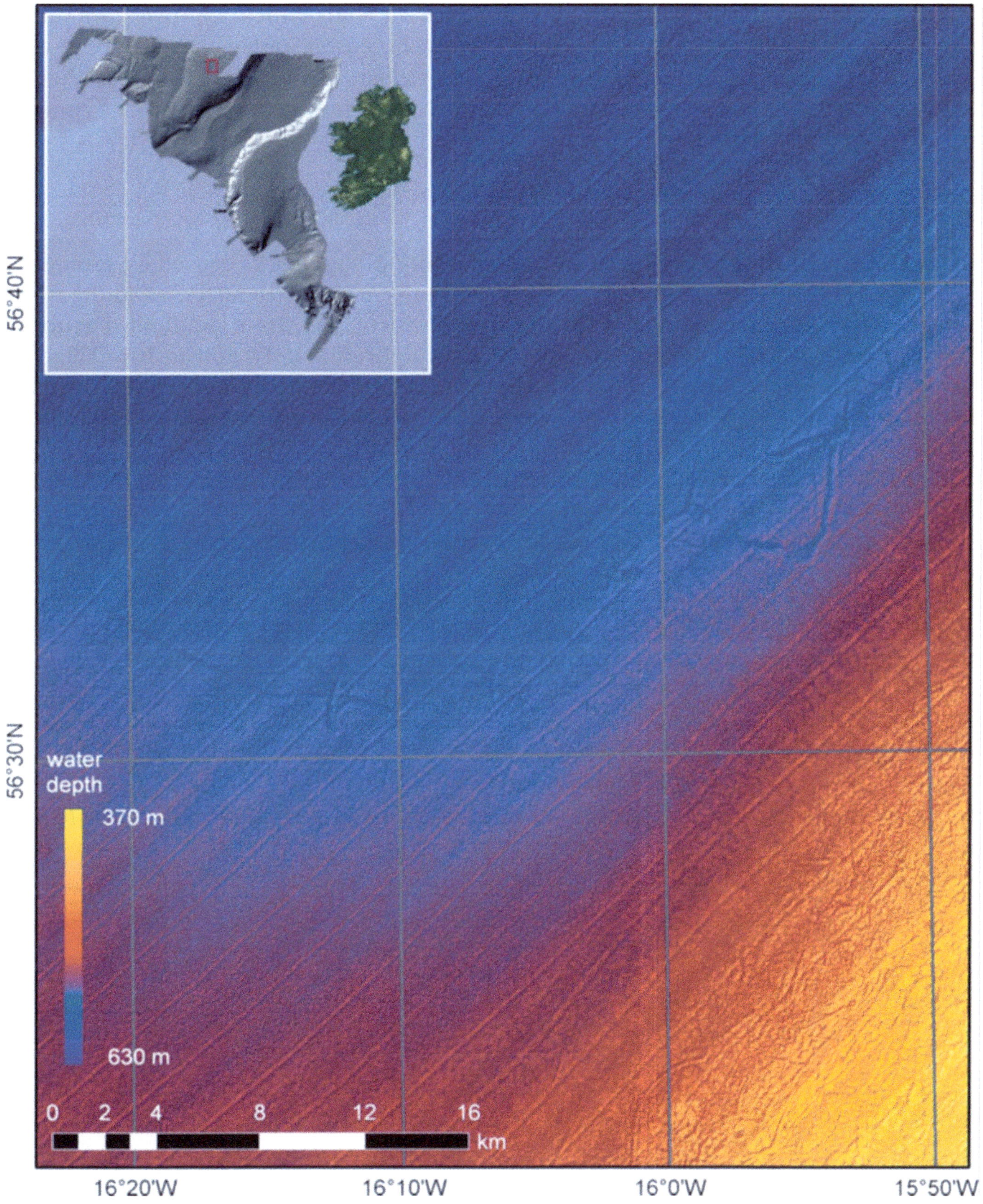

Fig. 11.4 **Iceberg ploughmarks on the Rockall Bank.** The deepest iceberg ploughmark occurs on the western slope of the Rockall Bank at 580 m water depth

mobile sediments fill the depressions and currents erode the levees.

11.1 Identification

Iceberg ploughmarks on the Irish seafloor were identified on the basis of hillshaded images with a 100 times amplified contrast. With this approach the shadows generated for the hillshade images remained the same size but the shadow intensity was much higher. As a result, small variations in bathymetry, which were usually overprinted by the changes of the general morphology, became enhanced and visible. A small, metre-deep ploughmark would not show up on a normal bathymetry map covering a depth range from 200 to 3,000 m bsl, nor would it show upon a normal unamplified hillshade relief. On the amplified image, however, the shadow of the ploughmark appears as a distinct dark line. In addition to iceberg ploughmarks, other sedimentary features such as sediment waves which often occur in the same depth range as the ploughmarks become more apparent and sometimes overprint them. For this reason, rather than mapping each ploughmark individually, the areas affected by icebergs are outlined.

11.2 Distribution

Iceberg ploughmarks caused by grounding icebergs, are mostly restricted to the upper regions of the Irish seafloor shallower than approximately 500 m bsl. They are particularly common on the western slopes of the Rockall and Porcupine Banks and in the Porcupine Seabight (Fig. 11.1).

Bibliography

Belderson, R. H., Kenyon, N. H., & Wilson, J. B. (1973). Iceberg plough marks in the northeast Atlantic. *Palaeogeography, Palaeoclimatology, Palaeoecology, 13* (3), 215–224.

Monteys X, Praeg D, Caloca S & Garcia-Gil S (2008). Iceberg keel marks on the Porcupine and Rockall Banks, NE Atlantic. AGU, 89(23), Jt. Assem. Suppl., OS24A-06.

Sejrup, H. P., Hjelstuen, B. O., Torbjørn Dahlgren, K. I., Haflidason, H., Kuijpers, A., Nygård, A., et al. (2005). Pleistocene glacial history of the NW European continental margin. *Marine and Petroleum Geology, 22*, 1111–1129.

Part III

Regional Atlas

Introduction

In this section, the Irish seabed deeper than 200 m water depth is subdivided into seven areas (Fig. 1). For each of the areas, an inventory of the seabed features occurring in that particular area is given (Table 1). In this way, the REGIONAL ATLAS supplements the THEMATIC ATLAS by showing all seabed features in their correct geo-referenced context.

Nomenclature: The names of large morphological features such as banks and troughs are written in CAPITALS, smaller morphological features, e.g. seamounts with official names, are written in normal font, and informal names that are in common usage, e.g. in scientific publications, but do not appear on existing seabed charts are written in *italics*.

The names used for seabed features throughout the publication are informal and generally those that have been used in publications or literature to date. An official protocol exists for the naming of offshore features, whereby names are ratified first by a national committee under the auspices of the Irish Hydrographic Office (Department of Transport) and thereafter by the International Hydrographic Organisation (IHO), before inclusion on official charts. Please note that the use of any such names in this publication does not imply any official recognition or support of the name used.

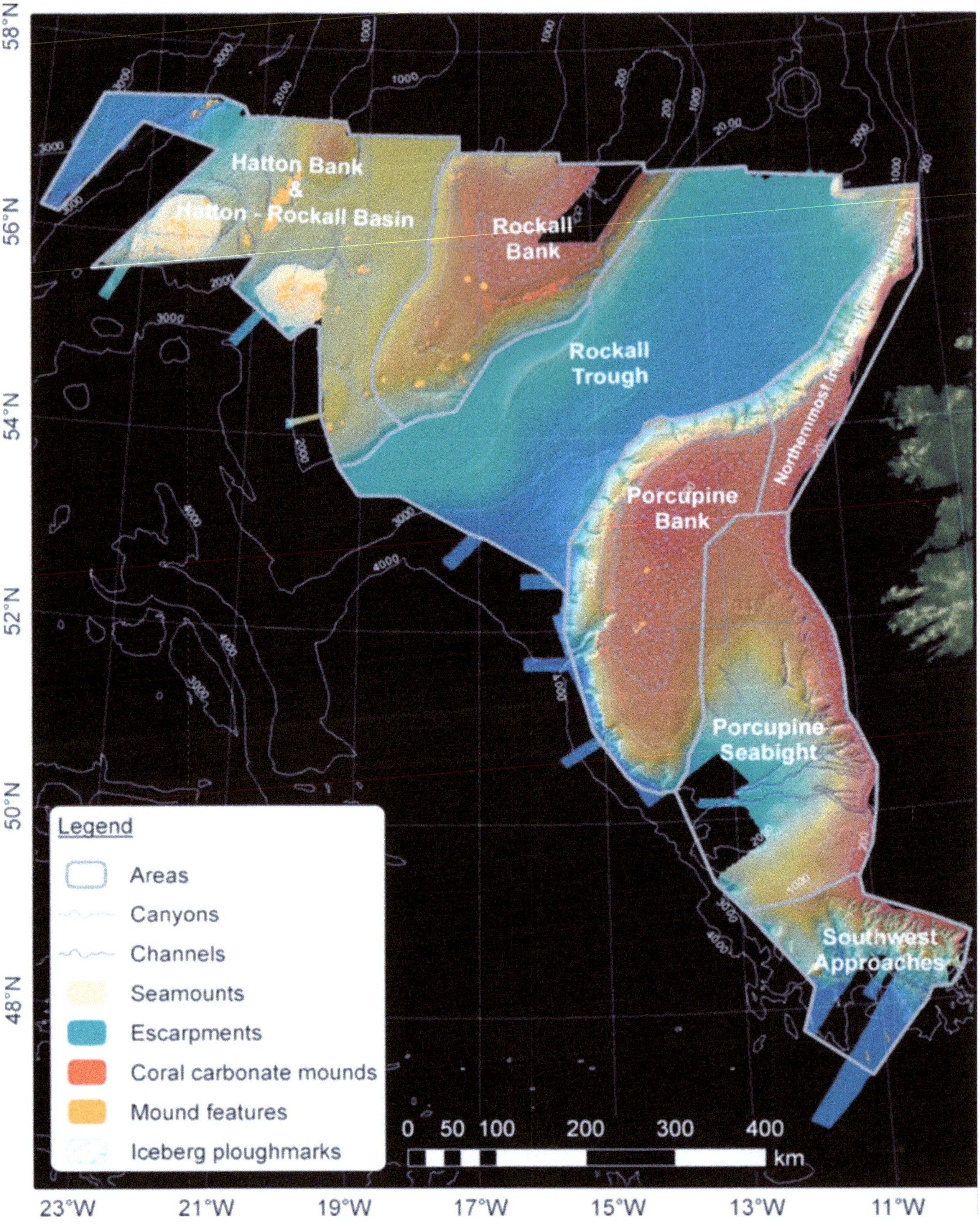

Fig. 1 **Areas of the Irish seabed.** For the regional atlas, the Irish seabed is sub-divided into seven areas representing the large geo-morphological features. Each area is individually presented in the chapters of Part III

Table 1 Number of features in the defined areas of the Irish seabed

	Canyons & Channels	Seamounts	Escarpments	Mound features	Coral mounds (ground-truthed*)
SWA pproaches	2	0	0	8	0
PorcupineS eabight	9	0	0	90	83
Porcupine Bank	20	0	47	463	404
NorthernmostI rish continentalm argin	18	0	13	8	0
Rockall Trough	0	1	8	8	0
RockallB ank	5	0	39	620	526
Hatton-RockallB asin & Hatton Bank	10	2	38	349	0
Total	64	3	146	1549	1013

* Mound features from groundtruthed coral carbonate mound provinces. Groundtruthed in this context means that these mound features are part of a mound province of which some mounds have been investigated with remotely operated vehicles, video cameras and box corers and confirmed as cold-water coral carbonate mounds

Southwest Approaches and Goban Spur 12

B. Dorschel, A.J. Wheeler, X. Monteys, and K. Verbruggen

The Southwest Approaches mark the southernmost extension of the Irish seabed and include the Irish part of the continental slope between the shelf area of the Celtic Sea in the east and the Bay of Biscay abyssal plain in the west (Fig. 12.1). This area is typified by numerous submarine canyons. During glacial periods, the Southwest Approaches represented the major pathway for sediments from northern Europe when their northward flow was blocked by the Fennoscandian and British-Irish ice sheets. Towards the northwest, the Southwest Approaches are bordered by the Goban Spur which also defines the south-eastern border of the Porcupine Seabight. The Goban Spur is a triangular extension of continental crust jutting out to the southwest. The Goban Spur was shaped during the opening of the Atlantic when a series of rift basins, partially separated by blocks of continental crust, were formed. Together, the Southwest Approaches and the Goban Spur cover 29,100 km^2 of the Irish seabed.

The most intriguing submarine features in this area are large submarine canyon systems. The most impressive example is the Whittard Canyon System, which, with its many feeding branches carved deeply into the Irish and UK continental slope, stretches for almost 300 km. The canyon systems of the Southwest Approaches are considered geologically active with seasonal pulses of suspended sediments channelled through them onto the deep abyssal plain. The Southwest Approaches represent a morphologically complex terrain and, even though not individually mapped, the very steep slopes contain countless escarpments along canyon walls and heads. The many branches of the submarine canyons form dendritic networks of almost artistic beauty resembling those of subaerial mountain river systems. The Southwest Approaches also contain six mound features of unknown origin which occur on the seabed near the Whittard Canyon.

B. Dorschel (✉) and A.J. Wheeler
Department of Geology & Environmental Research Institute, University College Cork, Cork, Ireland
e-mails: b.dorschel@ucc.ie; a.wheeler@ucc.ie

X. Monteys and K. Verbruggen
Geological Survey of Ireland, Beggars Bush, Haddington Road, Dublin 4, Ireland
e-mails: Xavier.Monteys@gsi.ie; Koen.Verbruggen@gsi.ie

B. Dorschel et al., *Atlas of the Deep-Water Seabed: Ireland*,
DOI 10.1007/978-90-481-9376-9_12, © Springer Science+Business Media B.V. 2010

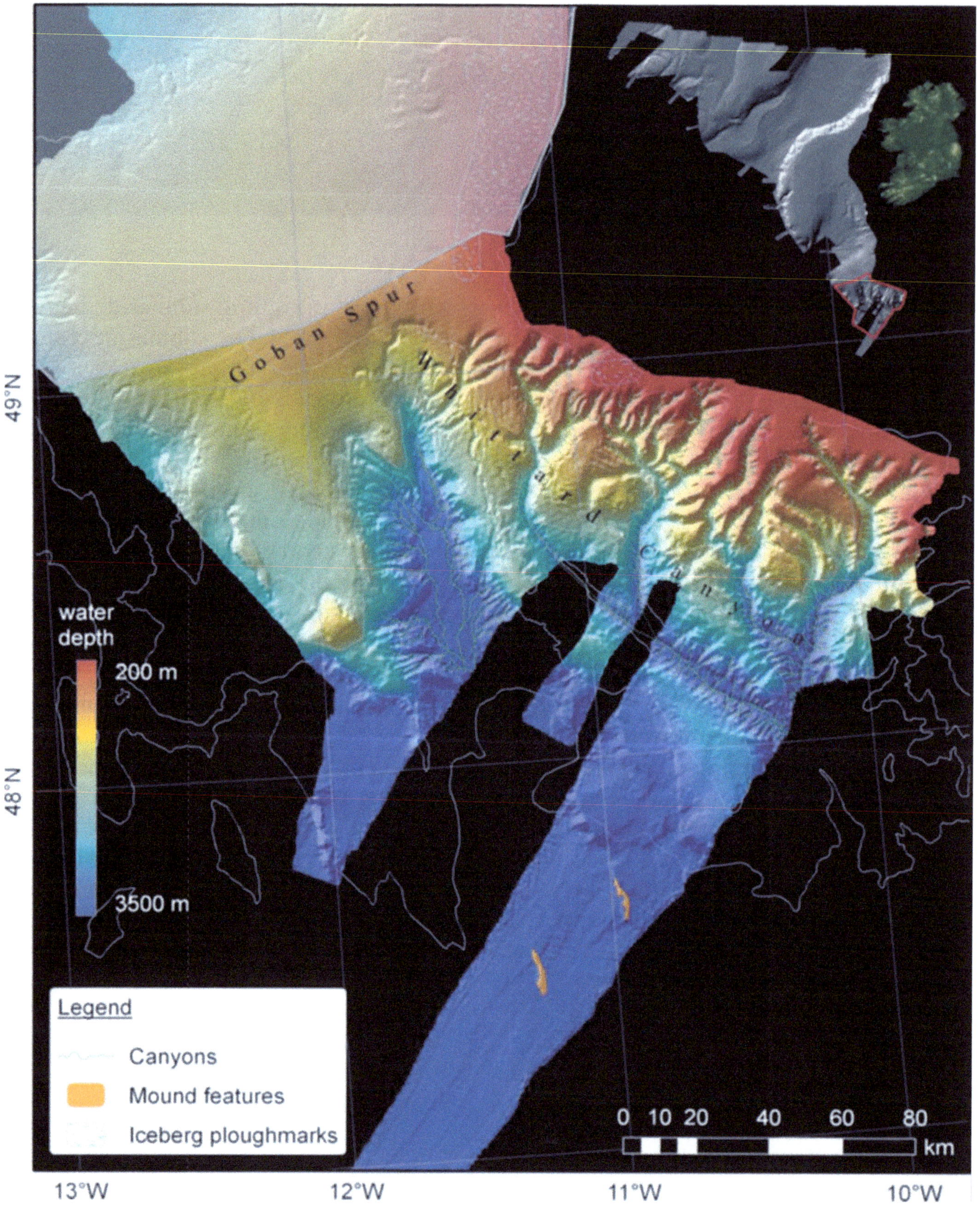

Fig. 12.1 **Southwest Approaches & Goban Spur.** The Irish part of the Southwest Approaches is characterised by the Whittard Canyon. The Whittard Canyon and the Southwest Approaches continue further to the southeast on the British and French seabed. In the north, the Goban Spur separates the Southwest Approaches from the Porcupine Seabight

Porcupine Seabight

13

B. Dorschel, A.J. Wheeler, X. Monteys, and K. Verbruggen

The Porcupine Seabight and the underlying Porcupine Basin developed as a failed rift structure when the Atlantic Ocean first started to open 250 million years ago. It is bordered by the Slyne Ridge in the north, the Porcupine Bank in the west and the Goban Spur in the south. The Porcupine Seabight opens to the southwest onto the Porcupine Abyssal Plain. Water depths in the Porcupine Seabight range from approximately 400 m in the north to 3,000 m at its mouth in the southwest. The northern border of the Porcupine Seabight is difficult to define with no clear break of slope but a gradual transition from the Seabight to the Porcupine Bank and Celtic Shelf. With the borders as presented here, the Porcupine Seabight covers an area of 67,100 km^2 (Fig. 13.1).

The meandering Gollum Channel system is the most prominent morphological feature in the Porcupine Seabight. It forms a large system of levees and channels that extend from the shelf edge on the eastern slope of the Porcupine Seabight, all the way through the central Seabight and down onto the Porcupine Abyssal Plain. It acts as the main pathway for focused sediment transport out of the Porcupine Seabight. In addition to the Gollum Channel system, several small channels occur in the northern Porcupine Seabight and a small unnamed canyon system can be found in deep waters at the southern tip of the Porcupine Seabight west of the Goban Spur. The formation of channels, rather than canyons, is typical for the gentle slope angle prevailing in the central and eastern Porcupine Seabight. On the northeast slope of the Porcupine Seabight, channels are in close proximity to coral carbonate mounds.

The Porcupine Seabight contains some of the best investigated cold-water coral carbonate mound provinces in the world. Three mound provinces are located in the Porcupine Seabight: the Belgica mound province on the eastern slope of the seabight, the Hovland mound province in the central northern Seabight and the Magellan mound province (a province of more than thousand mainly buried mounds) northwest, north and northeast of the Hovland mound province. The Belgica mound province also contains the Challenger Mound, the only cold-water coral carbonate mound that has ever been fully drilled from top to base. In addition to the well-studied coral carbonate mounds, eight unstudied mound features are located in the

B. Dorschel (✉) and A.J. Wheeler
Department of Geology & Environmental Research Institute, University College Cork, Cork, Ireland
e-mails: b.dorschel@ucc.ie; a.wheeler@ucc.ie

X. Monteys and K. Verbruggen
Geological Survey of Ireland, Beggars Bush, Haddington Road, Dublin 4, Ireland
e-mails: Xavier.Monteys@gsi.ie; Koen.Verbruggen@gsi.ie

B. Dorschel et al., *Atlas of the Deep-Water Seabed: Ireland*,
DOI 10.1007/978-90-481-9376-9_13, © Springer Science+Business Media B.V. 2010

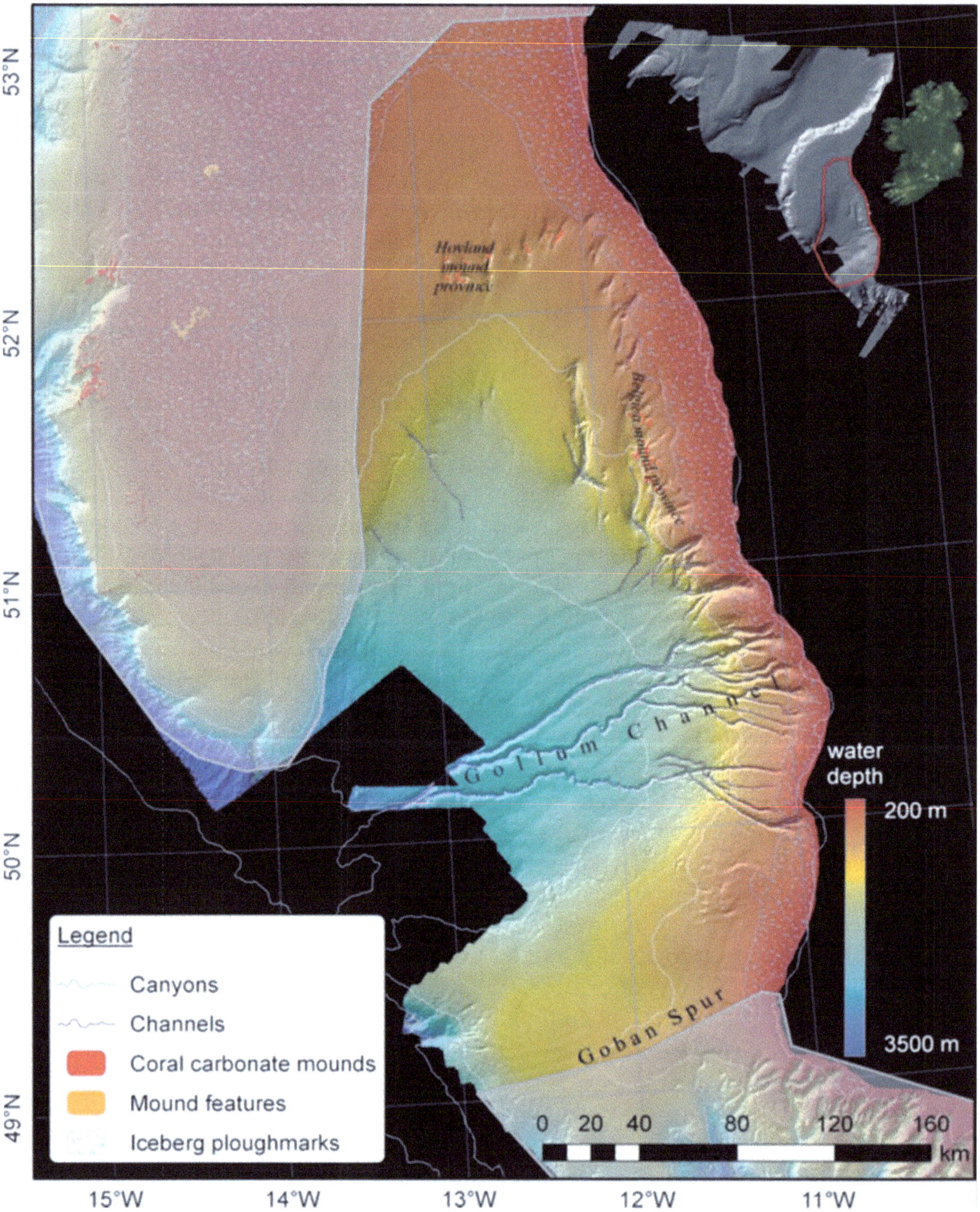

Fig. 13.1 **Porcupine Seabight.** The Porcupine Seabight is a tongue-shaped embayment in the Irish continental margin. Along with the Gollum Channel and numerous smaller channels systems, the Porcupine Seabight also contains probably the best studied coral carbonate mound provinces in the world

southern part of the Porcupine Seabight. Of these un-investigated mound features, two are associated with the channel heads of the Gollum Channel system (also known as the Kings Canyons) and eight form a small cluster north of the Goban Spur on the southeast slope of the Porcupine Seabight.

The areas of the Porcupine Seabight shallower than 500 m water depth are scarred by numerous iceberg ploughmarks. They are particularly common on the eastern and northern slopes of the Porcupine Seabight where individual ploughmarks can be traced for several kilometres zigzagging up the slopes. Towards the Gollum Channel, iceberg ploughmarks get successively overprinted by additional sedimentary features such as sediment waves and drifts.

Porcupine Bank

14

B. Dorschel, A.J. Wheeler, X. Monteys, and K. Verbruggen

The Porcupine Bank is a block of continental crust that became separated from the rest of the European continent by a failed rift during the opening phase of the North Atlantic, which began about 250 million years ago. The Porcupine Bank separates the Porcupine Seabight from the Rockall Trough. In the northeast, the Porcupine Bank is still connected with the Irish continental margin by the underlying structure of the Slyne Ridge. The summit of the Porcupine Bank is shallow lying at 145 m water depth and is generally broad and flat, although some structures from the underlying basement rocks can be seen emerging in places. The eastern slope, towards the Porcupine Seabight, is gentle whereas the southern, western and northern slopes towards the Rockall Trough are steep. Along the western and northern Porcupine Bank, the slope-break from the flat summit area onto the steep slopes occurs at a remarkably consistent water depth of approximately 450 m bsl and is generally marked by a prolonged escarpment. The western and northern slopes of the Porcupine Bank facing the Rockall Trough are characterised by irregularly spaced canyons and the south-western slope of the Porcupine Bank is especially steep and eroded (Fig. 14.1).

Canyon systems characterise the entire western and northern slopes facing the Rockall Trough and the southern part of the Porcupine Bank that faces the Porcupine Seabight. The most prominent canyon system is carved into the western slope of the Porcupine Bank and is known as the Porcupine Bank Canyon. This canyon is aligned with a major fracture in the oceanic plate to the west (the Charlie-Gibbs Fracture Zone) and may also be underlain by a major fault influencing the canyon's size and orientation. South of this canyon system, smaller canyons occur at 40–50 km intervals around the southern tip of the bank, while towards the north, canyons occur more closely spaced, usually less than 30 km apart. In total, 30 canyon systems are carved into the slopes of the Porcupine Bank. The relation between steep slopes and canyons, and gentle slopes and channels is clearly developed along the northern slope of the Porcupine Bank. There, the slope angles gradually decrease towards the Irish continental margin and the canyon systems gradually show more channel-like

B. Dorschel (✉) and A.J. Wheeler
Department of Geology & Environmental Research Institute, University College Cork, Cork, Ireland
e-mails: b.dorschel@ucc.ie; a.wheeler@ucc.ie

X. Monteys and K. Verbruggen
Geological Survey of Ireland, Beggars Bush, Haddington Road, Dublin 4, Ireland
e-mails: Xavier.Monteys@gsi.ie; Koen.Verbruggen@gsi.ie

B. Dorschel et al., *Atlas of the Deep-Water Seabed: Ireland*,
DOI 10.1007/978-90-481-9376-9_14, © Springer Science+Business Media B.V. 2010

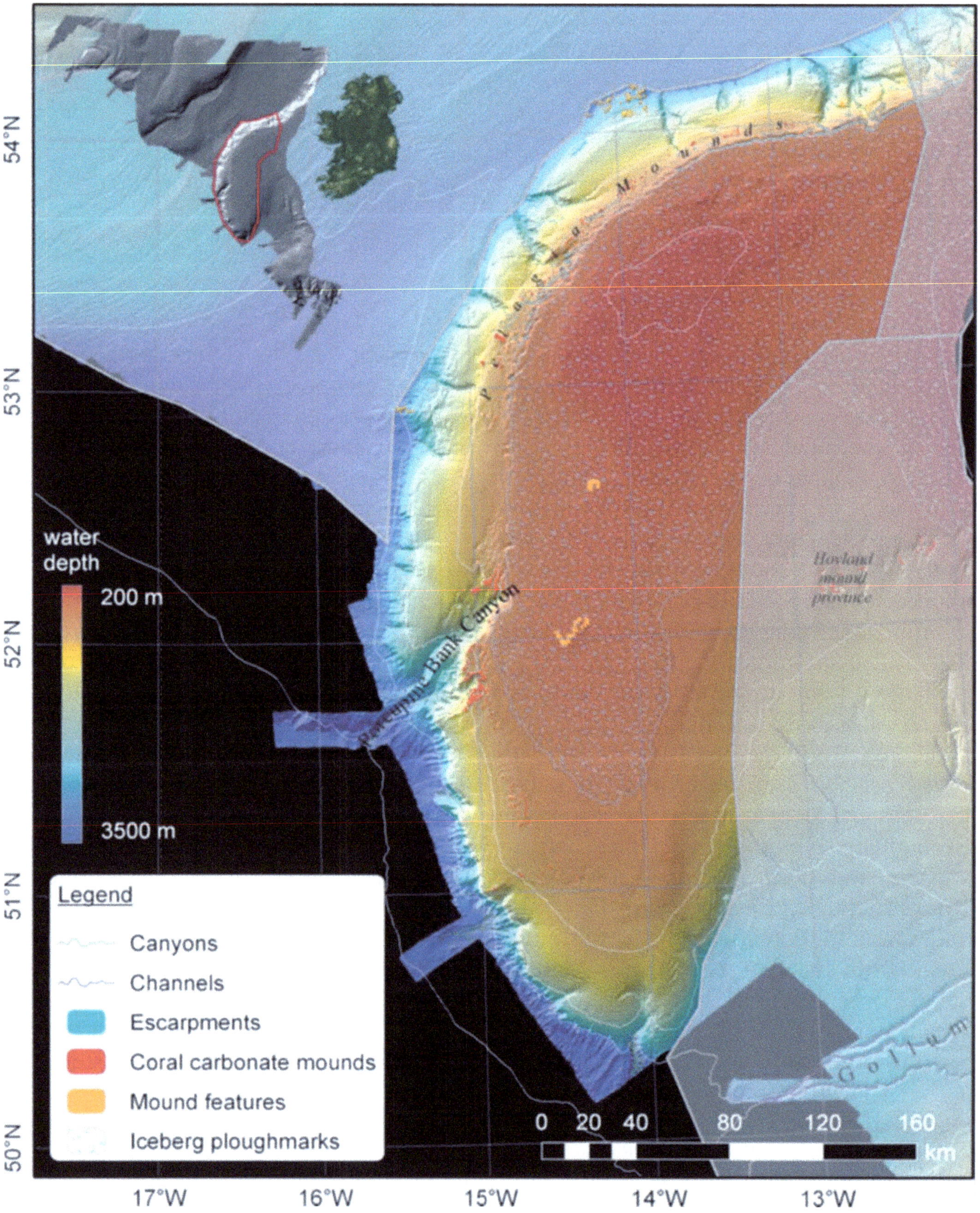

Fig. 14.1 **Porcupine Bank.** The Porcupine Bank is a continental block partly split off the Irish shelf by a failed rift during the opening of the Atlantic. As such, it separates the Porcupine Seabight from the Rockall Trough. It is characterised by gentle east and steep south, west and northwest slopes. Along the steep slopes of the Porcupine Bank, several canyons, escarpments and mounds occur while the flat shallow areas of the bank show widespread iceberg ploughmarks

features. In comparison with the canyon systems of the Southwest Approaches, the Porcupine Bank canyon systems are located in deeper water depth with canyon heads between 700 and 1,000 m water depth. The origin of these deep canyon heads is a mystery as the water depths do not coincide with the glacial sea level low-stands and are disconnected from direct sediment input from the shelf or bank top. It is speculated that the canyons may have been formed through erosion by tidal currents washing deep water-masses up and down the bank.

Escarpments can be found in many places along the western side of the Porcupine Bank. Many of them are associated with canyon heads but the majority occur as the result of strong erosive bottom currents. The most prominent escarpments strike slope-parallel between 450 and 650 m water depth and mark the slope break between the flat summit of the Porcupine Bank and the steep northwest slope. This water depth also coincides with regional water mass boundaries along which internal tides can result in locally amplified bottom currents which are strong enough to move and erode even pebble-sized particles.

At the northwest corner of the Porcupine Bank, an assemblage of mound features can be found in water depths between 2,100 and 2,400 m bsl. These mound features are of volcanic origin and represent 'mini seamounts'.

Provinces of coral carbonate mounds can also be found on the northern and western slopes of the Porcupine Bank. The so-called Pelagia Mounds are located on the northwest Porcupine Bank while the so-called Porcupine Bank Canyon Mounds can be found further south. The mounds on the Porcupine Bank occur between 1,500 and 500 m water depth as elongated mound chains, in association with canyon heads and on top of escarpments.

Furthermore, two clusters of mound-like features can be found on top of the Porcupine Bank. One of the features is a cone shaped feature. The other is a cluster of low mounds in a circular fashion. Both features were groundtruthed and are covered by gravel.

Widespread iceberg scouring occurs on the Porcupine Bank, especially in the soft sediments of the south-eastern Porcupine Bank where iceberg ploughmarks are well preserved at water depths shallower than 500 m bsl.

Northernmost Irish Continental Margin

15

B. Dorschel, A.J. Wheeler, X. Monteys, and K. Verbruggen

The northernmost Irish continental margin is the extension of the continental margin north of the Porcupine Bank and west of Ireland. As such, it forms the northern extension of the eastern margin of the Rockall Trough. In its southern and central part, the northernmost Irish continental margin is characterised by relatively steep slopes and numerous canyon systems. Only in the very northern part do slope angles gradually decrease and there the margin is more strongly influenced by across-shelf sediment transport from the now extensive shelf north of Ireland and west of Scotland (Fig. 15.1). This additional sediment input is reflected in the smoothed topography at this stretch of the Irish continental margin. In the south of the area, a relative shallow platform east of the margin is present where a thick sequence of sediment has infilled the northern Porcupine and Slyne Basins. This area has been the focus of intensive hydrocarbon exploration over the last few decades.

Like the western side of the Porcupine Bank, the northernmost Irish continental margin is characterised by across-slope submarine canyon systems. In the southern part of this area, a series of smaller canyons occur at ~5 km intervals while further to the north, larger canyon systems occur at ~10 km intervals. On the gentle slopes of the northern part of this area, no canyon systems are developed. This change in the nature of the continental margin is also evident in the distribution of escarpments. Pronounced escarpments continue at the same depth interval (450–650 m water depth) from the western Porcupine Bank onto the Irish continental margin. In the central part of the area, escarpments are solely associated with canyon systems. The same trend is true for the mound features, with mound features being restricted to the southern part of the area where a cluster of seven large mound features occur. Only one single mound is located further to the north. Due to the lack of groundtruthing and sub-bottom data, nothing is known about the origin of these mound features although they occur in the appropriate depth range to be considered as potential coral carbonate mounds.

B. Dorschel (✉) and A.J. Wheeler
Department of Geology & Environmental Research Institute, University College Cork, Cork, Ireland
e-mails: b.dorschel@ucc.ie; a.wheeler@ucc.ie

X. Monteys and K. Verbruggen
Geological Survey of Ireland, Beggars Bush, Haddington Road, Dublin 4, Ireland
e-mails: Xavier.Monteys@gsi.ie; Koen.Verbruggen@gsi.ie

B. Dorschel et al., *Atlas of the Deep-Water Seabed: Ireland*,
DOI 10.1007/978-90-481-9376-9_15, © Springer Science+Business Media B.V. 2010

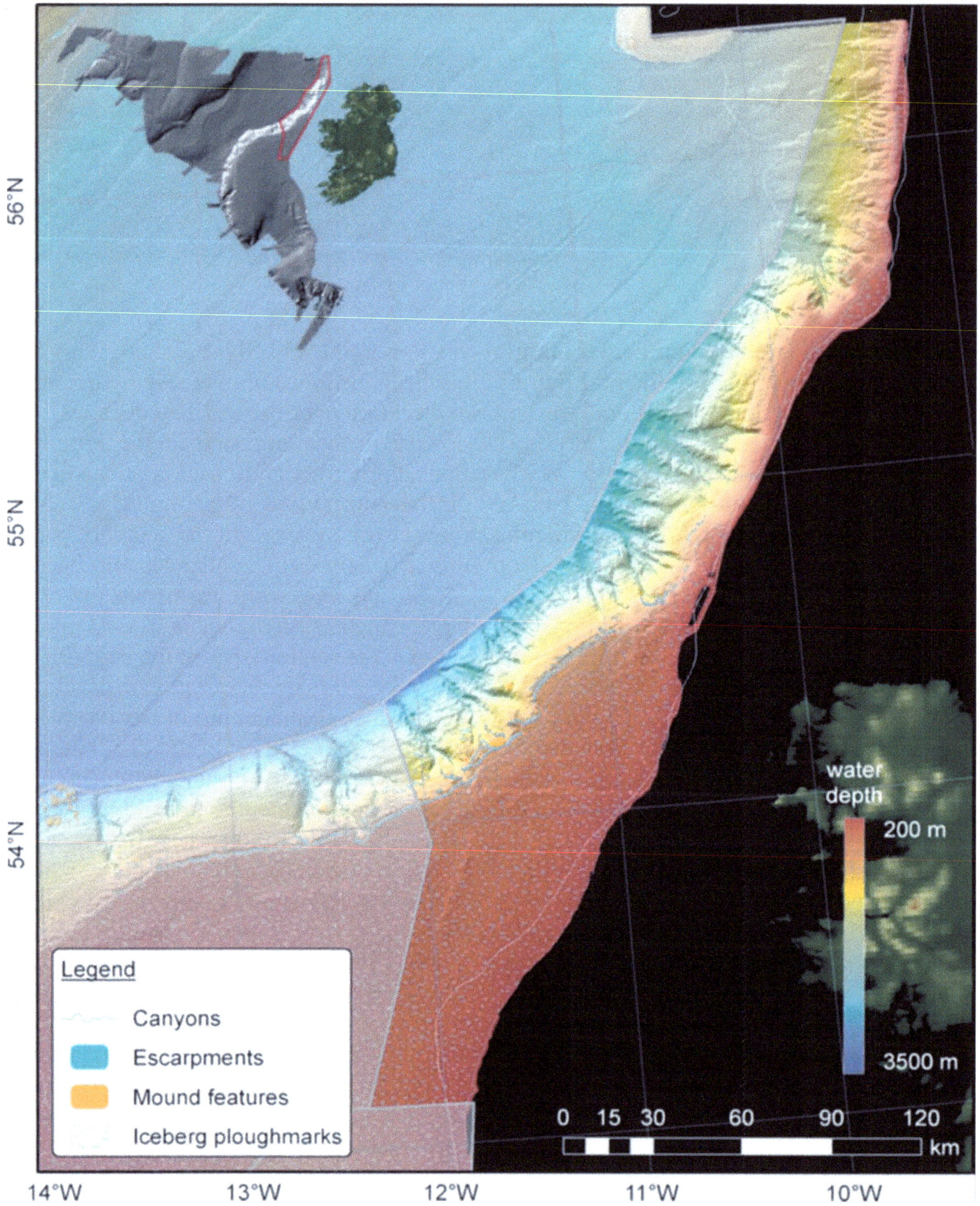

Fig. 15.1 Northernmost Irish continental margin. The northernmost Irish continental margin is the northward continuation of the Porcupine Bank margin. Along the southern stretch of the northernmost Irish continental margin, morphological features resemble those of the western slopes of the Porcupine Bank with numerous canyons and escarpments. Towards the north, the slopes of the northernmost Irish continental margin are less steep and more influenced by across-slope sedimentary processes

Iceberg ploughmarks are common on the southernmost stretch of the northernmost Irish continental margin, where it forms the northern extension of the Porcupine Bank. Towards the north, iceberg ploughmarks become less common. In this area they are probably overprinted or obscured by younger sedimentary features.

Rockall Trough

16

B. Dorschel, A.J. Wheeler, X. Monteys, and K. Verbruggen

The Rockall Trough is a southwest-northeast trending basin bordered by the Wyville Thomson Ridge in the far north (UK sector), Porcupine Bank and the northernmost Irish continental margin in the east and the Rockall Bank in the west. The Rockall Trough opens onto the Porcupine Abyssal Plain to the south (Fig. 16.1). Like the Porcupine Seabight, the Rockall Trough also evolved during the opening phase of the Atlantic starting 250 million years ago. However, despite small intrusions, no new oceanic crust has formed in the Rockall Trough leaving the Rockall and Porcupine Bank geologically connected to Ireland by thinned continental crust below the Rockall Basin. The Irish part of the Rockall Trough is 2,500 m deep in the north dropping to 3,500 m water depth in the south. The borders of the Rockall Trough with the Porcupine and Rockall Banks are clearly defined and can be identified by an abrupt increase in slope angle. Towards the north, these borders converge into a more gradually changing slope.

The Feni Ridge is a large drift sediment body located in the western part of the Rockall Trough. Due to the rotation of the Earth and the Coriolis force, moving masses are deflected *cum sole* (to the right) in the northern hemisphere. As a result, southerly flowing water entrained into the Rockall Trough on its journey from the Arctic does not flow down the centre but along the western margin. The Feni Drift is an accumulation of sediment heaped up by these strong currents. Smaller sedimentary features such as sediment waves and furrows are superimposed on the main drift sediment body. The second large morphological feature in the Rockall Trough is the Hebrides Terrace Seamount. This seamount is located at the northernmost extent of the Irish seabed, partly in Irish and partly in UK waters.

Escarpments can be found in the northwest part of the Irish Rockall Trough in water depths mainly below 1,800 m. These irregularly shaped escarpments most likely represent headwall scars from submarine slides.

B. Dorschel (✉) and A.J. Wheeler
Department of Geology & Environmental Research Institute, University College Cork, Cork, Ireland
e-mails: b.dorschel@ucc.ie; a.wheeler@ucc.ie

X. Monteys and K. Verbruggen
Geological Survey of Ireland, Beggars Bush, Haddington Road, Dublin 4, Ireland
e-mails: Xavier.Monteys@gsi.ie; Koen.Verbruggen@gsi.ie

B. Dorschel et al., *Atlas of the Deep-Water Seabed: Ireland*,
DOI 10.1007/978-90-481-9376-9_16, © Springer Science+Business Media B.V. 2010

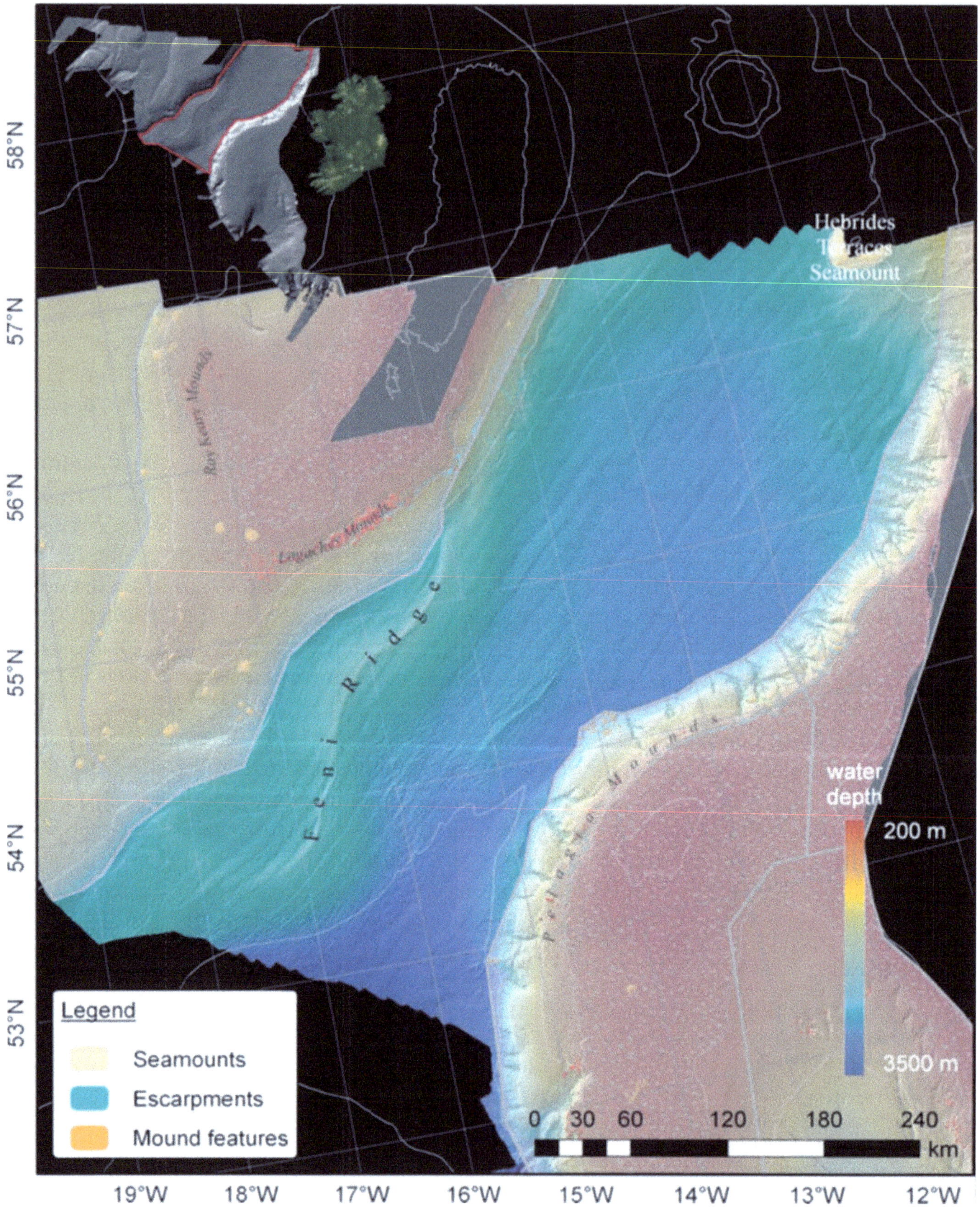

Fig. 16.1 Rockall Trough. The Rockall Trough formed as a failed rift during the opening phase of the Atlantic and is underlain by thinned continental crust and therefore geologically still connected with the European continental crust. Sedimentary features such as the Feni Ridge are the result of intense bottom currents in the Rockall Trough

Rockall Bank

17

B. Dorschel, A.J. Wheeler, X. Monteys, and K. Verbruggen

The Rockall Bank, to the west of the Rockall Trough, is another block of continental crust that became separated from the European continent during the opening phase of the Atlantic starting 250 million years ago. It is still connected to Europe by thinned continental crust below the Rockall Trough. The shallowest part of the Rockall Bank is the Rockall Islet, a small, bare rock island on the northern Rockall Bank that is permanently above sea level. Being separated from the Irish continental margin by the Rockall Trough, the Rockall Bank represents a relatively sediment starved area isolated from the input of terrestrial sediments by rivers and glaciers. Morphologically, the Rockall Bank is characterised by a generally steeper eastern slope and a generally gentler western slope (Fig. 17.1).

As a result of its isolation from sediment input, features created by sediment transport such as canyons and channels are uncommon on the Rockall Bank. Only three deep-seated canyons are carved into the eastern slope of Rockall Bank and two small channels occur on the western slope of Rockall Bank. Unlike sediment transport features, erosional features are common.

Escarpments are abundant on the eastern slope of the Rockall Bank towards the northern extent of the Irish seabed where they occur mainly in depth intervals from 450 to 750 m and 1,200 to 1,500 m water depth. On the western side of Rockall Bank, escarpments occur mainly in the depth interval from 750 to 900 m water depth also toward the northern extent of the Irish seabed. An additional five escarpments can also be found on the southern Rockall Bank.

The Rockall Bank hosts the majority, as well as the largest, mound features. In total 620 out of the total of 1,549 mapped mound features are located on the slopes of Rockall Bank. From these 620 mound features, 526 belong to two groundtruthed coral carbonate mound provinces, the so-called Logachev Mounds along the southeast slope of Rockall Bank and the so-called Ray Keary Mounds along the western slope of the Rockall Bank. Additional mound features of unknown origin are located on the southern tip of the Rockall Bank.

The Logachev mound province contains the largest examples of coral carbonate mounds in the world with some mounds

B. Dorschel (✉) and A.J. Wheeler
Department of Geology & Environmental Research Institute, University College Cork, Cork, Ireland
e-mails: b.dorschel@ucc.ie; a.wheeler@ucc.ie

X. Monteys and K. Verbruggen
Geological Survey of Ireland, Beggars Bush, Haddington Road, Dublin 4, Ireland
e-mails: Xavier.Monteys@gsi.ie; Koen.Verbruggen@gsi.ie

B. Dorschel et al., *Atlas of the Deep-Water Seabed: Ireland*,
DOI 10.1007/978-90-481-9376-9_17, © Springer Science+Business Media B.V. 2010

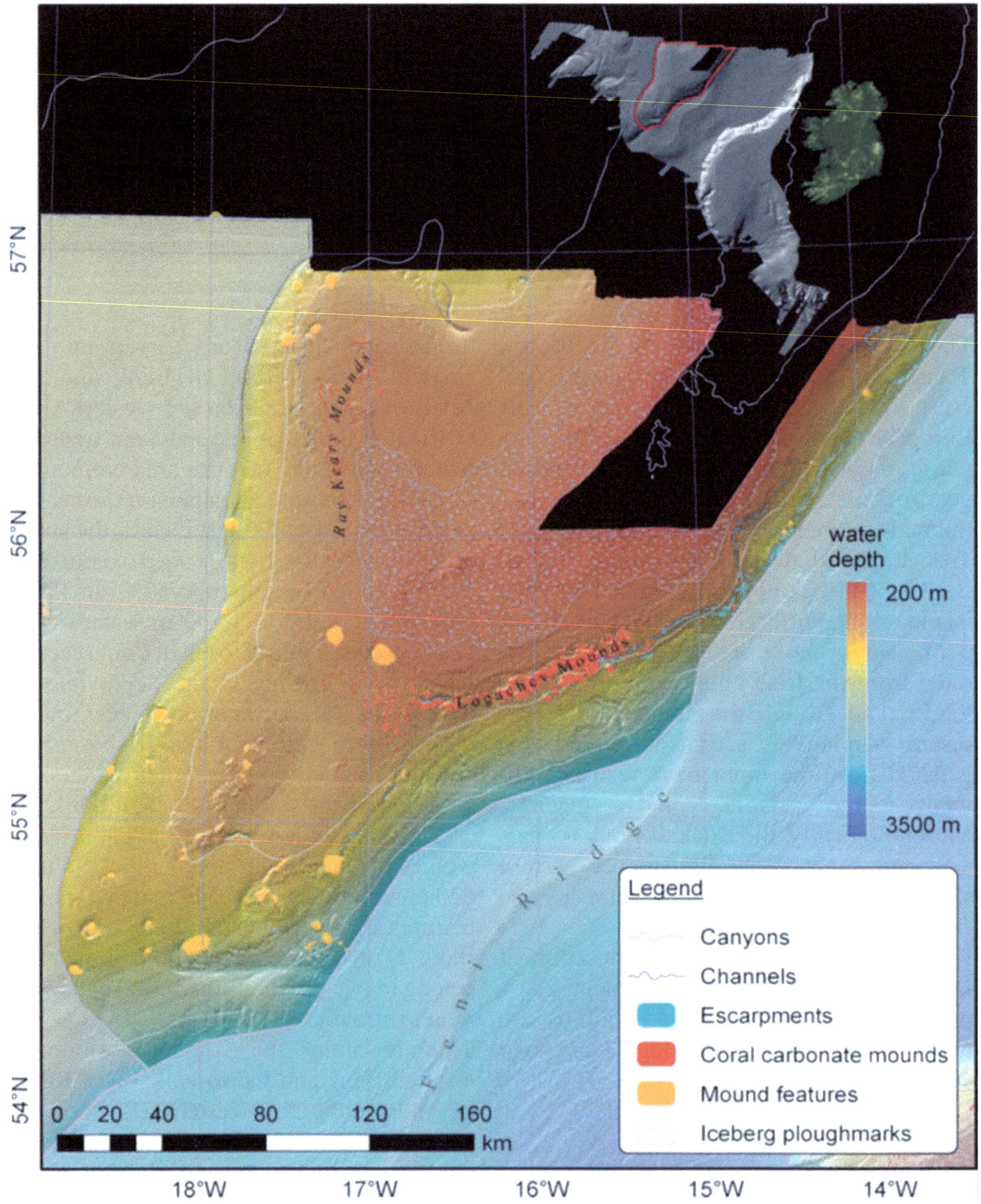

Fig. 17.1 **Rockall Bank.** The Rockall Bank is a block of continental crust separated from Porcupine Bank by a failed rift. Due to its isolated position, the input of terrigenous sediment on the Rockall Bank is relatively low. The eastern slopes of the Rockall Bank are especially characterised by erosion rather than deposition. The largest coral carbonate mounds known worldwide can be found on the southeast slope of the Rockall Bank

measuring over 500 m from the deepest part of the base to the summit. The Logachev Mounds are complex multi-summitted features often displaying vague slope perpendicular orientations. As opposed to the large Logachev Mounds, the Ray Keary Mounds on the western Rockall Bank comprise almost 300 small mounds with the largest rising 130 m above the surrounding seabed, although 90% of these mound are actually smaller than 45 m. The remaining mound features unfortunately lack any groundtruthing information and may be more coral carbonate mounds, mini-seamounts or rock outcrops.

Like all shallow areas of the Irish seafloor, the upper Rockall Bank has been affected by iceberg scouring in the past. The deepest iceberg ploughmarks discovered on the Irish seabed are located on the western slope of the Rockall Bank at 580 m water depth.

Hatton-Rockall Basin & Hatton Bank

18

B. Dorschel, A.J. Wheeler, X. Monteys, and K. Verbruggen

The Hatton-Rockall Basin and the Hatton Bank are the names given to the area of the designated Irish seabed furthest away from dry land. This area reaches all the way to the western slopes of the Iceland Basin. The whole area of the Hatton-Rockall Basin and the Hatton Bank represents the westward extension of the block of continental crust to which the Rockall Bank also belongs. It is therefore characterised by shallower water depths with even the deepest parts of the Hatton-Rockall Basin not reaching below 2,000 m water depth, considerably shallower than the Rockall Trough (Fig. 18.1). Due to its remote location, the area of the Hatton-Rockall Basin and the Hatton Bank has rarely been targeted by research cruises, however it was fully mapped by the INSS and PAD programmes. As a result of its remoteness, very little groundtruthing data is available for the morphological features in this area.

B. Dorschel (✉) and A.J. Wheeler
Department of Geology & Environmental Research Institute, University College Cork, Cork, Ireland
e-mails: b.dorschel@ucc.ie; a.wheeler@ucc.ie

X. Monteys and K. Verbruggen
Geological Survey of Ireland, Beggars Bush, Haddington Road, Dublin 4, Ireland
e-mails: Xavier.Monteys@gsi.ie; Koen.Verbruggen@gsi.ie

Submarine canyons do not occur in the Hatton-Rockall Basin and the Hatton Bank area, although several channel systems indicate the major sediment pathways in the area and highlight the general flow pattern around morphological obstacles such as seamounts. At the very western extension, the bathymetric data cover part of the Maury Channel system.

Two prominent seamounts occur in this area south of the Hatton Bank. They rise more than 1,000 m above the surrounding seafloor and as such can be termed "large seamounts".

As the entire area of the Hatton-Rockall Basin and Hatton Bank are characterised by gently inclined seafloor, escarpments are not very common. The identified escarpments are therefore mainly associated with the steep flanks of the seamounts and Hatton Bank.

Three hundred and forty-nine mound features are located in this area. These features occur superimposed on the seamounts, or as ridges and clusters. The origin of these features is mainly unknown due to the lack of groundtruthing data although it is likely that some of the mound features represent coral carbonate mounds as they are located within the depth range of these mounds.

With the shallowest seafloor in this area more than 750 m deep, the area is too deep to possess iceberg ploughmarks.

B. Dorschel et al., *Atlas of the Deep-Water Seabed: Ireland*,
DOI 10.1007/978-90-481-9376-9_18, © Springer Science+Business Media B.V. 2010

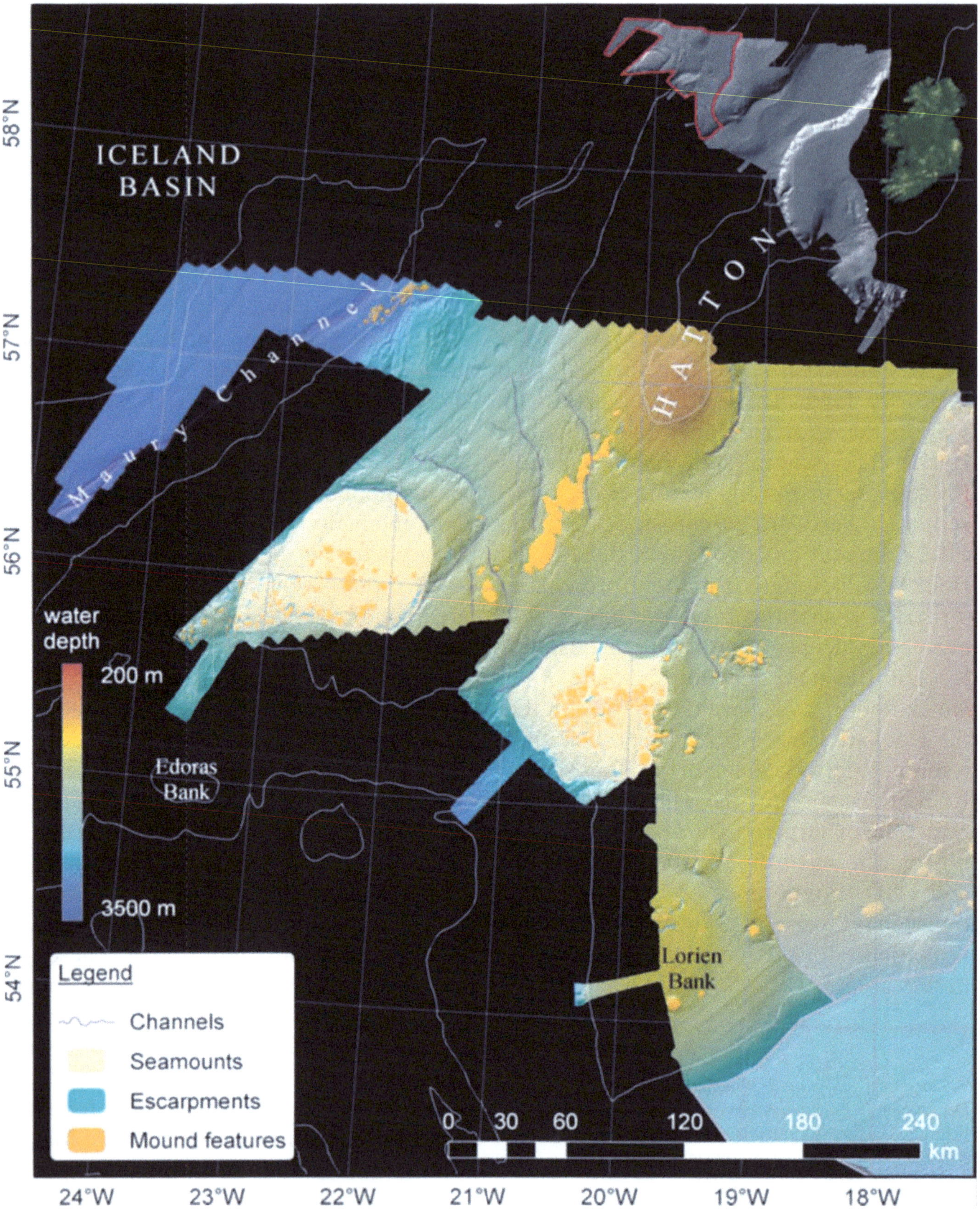

Fig. 18.1 Hatton-Rockall Basin & Hatton Bank. The Hatton-Rockall Basin and Hatton Bank mark the westernmost extension of the Irish seabed. As such, it is the westward extension of the block of continental crust forming the Rockall Bank. Due to its remote location very little is known about the morphological features on this part of the Irish seafloor

Part IV

Going Deeper

19 A Guide to the Digital Atlas

B. Dorschel, A.J. Wheeler, X. Monteys, and K. Verbruggen

The main objective of this atlas is to take the reader on a journey to largely unexplored and fascinating areas of Ireland's deep ocean. However, this is not just a colourful book, this atlas is much more...

This atlas is only a taster. The hard copy of the atlas you are holding in your hands, or the e-book you are looking at, is only the topmost layer of an archive of datasets. To produce an atlas like this, multiple layers of information are piled on top of each other generating a condensed information package that is presented here in the form of maps (Fig. 19.1). However, this is not a once-off product, the information layers can be removed, recombined, changed and modified. In this way, the maps we present are only possible products and examples of what can be done with the INSS dataset. The following sections of this chapter provide information on how to access the data from which this atlas was made and where to find additional marine data sets to further explore the deep ocean seabed.

There are many uses for the INSS dataset. For anybody working on, or with, the seafloor of the deep-sea, good quality, high resolution bathymetry data are essential for planning purposes or for conducting surveys. Marine researchers, for example, require detailed bathymetry maps for surveys with remotely-operated vehicles or for sampling campaigns. Fishermen require seabed maps to locate the type of terrain that may support different types of benthic fisheries and gauge how deep to tow, or where not to tow, their nets. Marine engineers require maps for planning cable or pipeline routes or anchoring drilling platforms on a stable seabed. For people generally interested in the high seas, this atlas and the datasets it is based on, simply make the invisible visible.

Having reached the last chapter of this seabed atlas, we hope that we have convinced you that the deep oceans are fascinating places. Hidden under the waves are mountain ranges, cliffs and volcanoes. Even though dark and cold, the deep seabed is inhabited by numerous, often beautiful, animals. Maybe, when you are next standing the at seashore looking over the vastness of the sea, you may better appreciate that there is also a still largely unexplored but now thankfully mapped, seabed spreading out in front of you as spectacular as any scenery you will find on land.

B. Dorschel (✉) and A.J. Wheeler
Department of Geology & Environmental Research Institute, University College Cork, Cork, Ireland
e-mails: b.dorschel@ucc.ie; a.wheeler@ucc.ie

X. Monteys and K. Verbruggen
Geological Survey of Ireland, Beggars Bush, Haddington Road, Dublin 4, Ireland
e-mails: Xavier.Monteys@gsi.ie; Koen.Verbruggen@gsi.ie

B. Dorschel et al., *Atlas of the Deep-Water Seabed: Ireland*,
DOI 10.1007/978-90-481-9376-9_19, © Springer Science+Business Media B.V. 2010

Fig. 19.1 Layered Maps. Even though flat, maps are often composed of many layers of information. Only with layers of information stacked on top of each other, it is possible to achieve the density of information provided in this atlas

19.1 A Matter of Perspective

When presenting geographic data, the perspective and the viewpoint are of crucial importance. It is for example impossible to identify a mound on a map of the entire north Atlantic. In the same way, photographs of the seabed only show several square metres leaving everything outside the field of view in the dark. By presenting high resolution overview maps in combination with detail maps and sometimes very detailed photos, we intend to illustrate the seabed and morphological seabed features from different perspectives (Fig. 19.2).

The large scale geological processes and their products become apparent on the overview maps. Combined with the underlying geology (Chapter 4) it is possible to imagine the opening of the Atlantic and the breaking of the old continent of Pangeae into the various blocks that now form the banks.

To illustrate smaller scaled processes, like the transport of sediments across the continental slope, erosion and the build-up of biogenic structures such as coral carbonate mounds, more detailed maps are required. Only high resolution bathymetry data sets like for example the INSS data provide enough detail to illustrate canyon and channel systems, escarpments and mounds. These data allow a view on the submarine landscape from the perspective that we are familiar with in the terrestrial realm.

But what would be an atlas without showing the seabed as it actually looks like. The development of deep-sea cameras and remotely-operated vehicles over the past decades provided the necessary tools to get a glance at the seafloor. Due to the technical advances, scientists are now in a position to record visual photographic and video imagery of the seafloor and even to stop and gain a closer, more detailed look.

Cold-water coral carbonate mounds illustrate how these changes of perspective results in a more holistic understanding (Fig. 19.2). Looking at the entire Irish seabed shows that cold-water coral carbonate mounds occur in distinct provinces in a defined water depth range on the upper continental slope. A closer look identifies the mounds in individual provinces characterised by similar morphological properties. But only the groundtruthing with video data and seafloor sampling clearly identifies mound features as actual cold-water coral carbonate mounds with thickets of cold-water corals growing on them. But how does a mounds function? To understand this, it is necessary to look at the organism building these mounds. With remotely operated vehicles and camera lander systems it is now possible to zoom right in onto individual corals and to record their behaviour and physiological processes where the corals live. And in a feedback process, information on the life of corals is used to understand how the mounds actually formed which again is linked to the environmental conditions influencing the corals and therefore controlling mound distribution.

The change of perspective broadens the view. We hope that by presenting the seabed and its morphological structures from various perspectives we have provided a new and hopefully exciting view on the deep seabed of Ireland.

Fig. 19.2 **A matter of perspective.** In order to gain a comprehensive picture of the seabed and the processes active on the seabed, it is necessary to look from different perspectives. Sometimes, the very detailed views provide clues that help to better understand large morphological features (ROV images courtesy of MARUM – Center for Marine Environmental Sciences, University of Bremen)

19.2 Accessing the On-line Data Sets Behind the Atlas

The actual data points used to produce the gridded bathymetry, on which the seabed atlas is based, can be accessed and downloaded free of charge from the web. Different maps of the seabed, along with graphical representations of items such as shipwrecks, sample locations and physical habitat maps can also be viewed and downloaded for free. All of this data can be accessed from the websites of either the Geological Survey of Ireland, the Marine Institute, or the website of the INFOMAR Project. See the web links below, all of which were valid at the time this publication went to press.

- Geological Survey of Ireland, (Department of Communications, Energy & Natural Resources), www.gsi.ie
- Marine Institute, www.marine.ie
- INFOMAR Project, www.infomar.ie
- Web mapping at http://spatial.dcenr.gov.ie/imf/imf.jsp?site=INFOMAR
- Downloaded data at https://jetstream.gsi.ie/iwdds/index.html

19.3 Useful Links and Contacts

Due to technical advances, research activities on the deep seabed have increased almost exponentially over the past decades. As a result, numerous sources of information and data on the deep seabed have become available. Here we present a small list of the sources of information on the deep seabed with a focus on the northeast Atlantic. This list is by no means complete but it can be used as a first step for a data search:

- Bedford Institute of Oceanography, www.bio.gc.ca
- Census of Marine Life, www.coml.org
- Environmental Research Institute, www.ucc.ie/en/eri
- EU-SEASED – Seabed samples from the Ocean Basins and European Continental Waters, www.eu-seased.net
- European Directory of Marine Environmental Data sets (EDMED), www.seadatanet.org/metadata/edmed
- FishBase - A Global Information System on Fishes, www.fishbase.org
- General Bathymetric Chart of the Oceans (GEBCO), www.gebco.net
- Geological Survey of Ireland, www.gsi.ie
- Goddard Institute for Space Studies, www.giss.nasa.gov
- IFM-GEOMAR – Leibniz-Institut für Meereswissenschaften an der Christian-Albrechts Universität zu Kiel, www.ifm-geomar.de
- IFREMER – Institut Français de Recherche pour l'Exploitation de la Mer, www.ifremer.fr
- INtegrated Mapping FOr the Sustainable Development of Ireland's MArine Resource (INFOMAR), www.infomar.ie
- Irish National Seabed Survey (INSS), www.gsi.ie/Programmes/INFOMAR+Marine+Survey/INFOMAR+Overview.htm
- Lophelia.org – Cold-water, deep-sea and deep-water coral resource, www.lophelia.org
- Marine Institute of Ireland, www.marine.ie
- MARUM – Center for Marine Environmental Sciences, www.marum.de
- MESH – Development of a framework for Mapping European Seabed Habitats, www.searchmesh.net
- National Geographic Data Center, www.ngdc.noaa.gov
- National Oceanic and Atmospheric Administration, www.noaa.gov
- National Oceanography Centre, Southampton, www.noc.soton.ac.uk
- National Parks and Wildlife Service (NWPS), www.npws.ie
- National University of Ireland, Galway, www.nuigalway.ie
- NIOZ – Royal Netherlands Institute for Sea Research, www.nioz.nl
- OBIS – Ocean Biogeographic Information System, www.iobis.org
- OSPAR Commission, www.ospar.org
- PANGAEA Publishing Network for Geoscientific & Environmental Data, www.pangaea.de
- Petroleum Affairs Division, www.pad.ie
- Renard Centre of Marine Geology, www.rcmg.ugent.be
- Scottish Association for Marine Science, www.sams.ac.uk
- Senckenberg by the Sea
- UK DECC Strategic Environmental Assessment, www.offshore-sea.org.uk/site/scripts/sea_archive.php
- University College Cork, www.ucc.ie
- Universtity College Dublin, www.ucd.ie

Glossary

Abyssal plain Flat or gently sloping seabed usually found between 3,000 and 6,000 m water depth

Aragonite Orthorhombic calcium carbonate mineral ($CaCO_3$); one of the two common, naturally occurring calcium carbonate minerals (see calcite)

Artefact In the context of this atlas, any error in bathymetric information introduced by the equipment or during processing

Bank Generic term for a large or small elevated topographic seafloor feature

Basement rocks Crystalline rock below sedimentary cover

Basin Generic term for large (regional scale) seafloor depression

Bathymetric data Geographic information on the depth of oceans and lakes below sea-level or lake level respectively

Bedrock Consolidated rock below broken and weathered rock or sediment cover

Bio-erosion Erosion of hard substrates and skeletal material by living organisms (in the marine environment)

Biogenic seabed features Seabed features produced or built by living organisms

Biological "hot spots" Areas of high species abundance and/or diversity and biological activity

Boom-and-bust exploitation Unsustainable and rapid extraction of natural resources

Bottom current Moving water mass at the seabed

Bottom water Water mass in the immediate vicinity of the seabed

British-Irish ice sheet Ice mass formed on Ireland and Britain during glacial periods

Calcite Trigonal-rhombohedral calcium carbonate mineral ($CaCO_3$); one of the two common, naturally occurring calcium carbonate minerals (see aragonite)

Canyon A deep and narrow valley with steep sides

Canyon head The upper termination of a canyon

Cell In the context of this atlas, the smallest geographical (or geo-referenced) unit with a unique value in a raster data set

Cell size In the context of this atlas, the dimensions of a cell in a raster data set

Cemented sediment Sediment consolidated/lithified by minerals in the interstices between the original sediment particles

Channel Generally, the bed of a river or stream. In the context of this atlas, the preferred pathway of suspended sediment flows

Cold-water coral Corals without symbionts that can live in cold and often deep waters

Cold-water coral carbonate mound Biogenic seabed feature formed due to repeated cold-water coral reef formation

Competent In the context of this atlas, rock resistant to plastic deformation

Conservative properties In seawater, properties of a water mass that only change when the water is in contact with the atmosphere or due to mixing with other water masses. The most prominent conservative properties are temperature and salinity

Consolidated sediment Sediment hardened by compaction and newly formed interstitial minerals (see cemented sediment)

Continental block Large (regional scale) segment of continental crust

Continental crust Earth crust found below continents which is mainly granitic in composition. Chemically distinct from the basaltic crust that underlies the oceans

Continental margin Zone of submerged continental crust extending from the shoreline to the beginning of oceanic crust

Continental shelf Seabed found from the shore to the shelf break. Approximately 0–200 m water depth (see also Chapter 1.1)

Continental slope Seabed found from the shelf break to the continental rise. Approximately 200 to 2,000 m water depth (see also Chapter 1.1)

Contour currents Bottom water currents flowing parallel to the bathymetric contours of the seabed

Coral carbonate mound See cold-water coral carbonate mound

Coral carbonate mound province A grouping of coral carbonate mounds in one area often with similar morphologies

Coral thicket Dense accumulation of coral colonies

Coriolis force A force deflecting objects (e.g. water, air) moving over the earth. It acts at right angles to the direction of motion

Current meter mooring Array of underwater analytical equipment to measure water motion

Data density Number of data points per area

Digital elevation model (DEM) A digital representation of the surface (in this context the seabed) topography

Down-welling The sinking of a higher density water mass below a lower density water mass

Drift sediment Sediments transported and accumulated by bottom currents

Dyke Vertical or steeply inclined sheet intrusion of magma

Ecosystem The biological community and its environment forming an ecological unit

Eddy A circular current

Endemic species A species only occurring in a defined geographic area

Environmental conditions Aspects of an environment (e.g temperature, salinity, current speed) that influence an ecosystem

Erosion Removal of sediments and rocks

Erosional feature Feature formed by past or ongoing erosion

Erosional scarp A lineament of cliffs produced by erosion

Erosive bottom current Bottom currents strong enough to remove sediment

Erratic boulder Metre-scale rock fragment transported by a glacier or iceberg from its original place

Escarpment A long cliff. In the context of this atlas, a lineament of steep slope where the seabed drops in several often vertical steps

Failed rift Initial rift (see rift) that stopped without the formation of oceanic crust

Fault A fracture of the earth crust where both sides have moved relative to each other

Fennoscandian Geological term for the area of Scandinavia, Finland and the Kola Peninsula

Geographic data Information relating to a position on the Earth's surface

Geo-morphological structure Morphological structure formed as the result of geological processes (e.g. a volcano)

Geo-statistical analyses Statistical analyses on the spatial distribution of features on the Earth's surface

Glacial Cold geological time interval when glaciers were more extensive than today

Glacial maximum Time of maximum glacier expansion during a glacial

Global positioning systems (GPS) Satellite based system to determine positions on the earth

Groundtruthing To verify interpretations of remote imaging data with sample and video data

Habitat Place or environment where an organism lives

Hard substrate Any type of solid ground (as opposed to soft substrate such as mobile sediments)

Hardground Lithified sediment with borings indicative of periods of non-deposition and erosion

Headwall In the context of this atlas, the cliff marking the highest extend of a submarine landslide

Hemipelagic sediment Sediment containing land derived components as well as parts of (mainly) planktonic organisms

High resolution data In the context of this atlas, geographic information resolving fine details of the seabed

Hillshaded image 2D image with shadows generating a 3D effect

Hinterland Area inland from the coast

Hydrographic environment Commonality of water current parameters affecting an area of seabed

Hydrological network analyses Estimation of the theoretical flow path of a parcel of water down a slope

Iceberg ploughmark Impact mark that a grounding iceberg leaves in the soft sediments on the seabed

Iceberg scouring Iceberg ploughing through soft sediments on the seabed

Incompetent In the context of this atlas, loosely or un-bound (sediment)

Internal tide Large scale waves generated along water-mass boundaries

Intrusion Magma intruded into other rock formations

Irish continental margin Westward extension of the Irish continental crust below the Atlantic (see also Chapter 1.1)

Lander system Frame mounted analytical instrument array deploy on the seabed

Levee Proximal channel-parallel ridge

Lithified Turned to stone (lithification of sediments)

Marine A general term of or relating to the sea

Marine sediment Sediments formed in the oceans as opposed to terrestrial sediments formed on land

Matrix In the context of this atlas, fine-grained sedimentary groundmass in which significantly larger lithic or biogenic clasts are embedded

Meander Winding path of a channel

Meridional currents Poleward flowing currents (in north-south or south-north directional, parallel to the meridians)

Mid-ocean ridge Submarine mountain range found at constructive plate boundaries where new oceanic crust is formed

Mini-seamount Small submarine volcanoes elevating less then 1.000 m above the adjacent seabed (see also seamount)

Mound base In the context of this atlas, the contact of the mound slope with the adjacent seabed

Mound feature Morphological structure elevating tens to hundreds of metres above the adjacent seabed

Non-conservative properties In seawater, properties of a water mass that can be changed by processes other than mixing with other water masses (e.g. nutrient content, CO_2 concentration)

Ocean basin Large geologic basin below sea level, underlain by oceanic crust

Ocean circulation Density defined flow pattern of different water masses at different water depth through the world's oceans

Ocean stratification Vertical change in properties through the water column due to the presence of water masses with different temperatures and salinities and hence densities

Ocean structure Vertical distribution of water masses

Oceanic crust Earth crust found below oceans which is mainly basaltic in composition. Chemically distinct from the granitic crust that underlies the continents

Organic matter Carbon compounds produced by organisms

Outcropping bedrock Bedrock exposed at the seabed

Pangea Super-continent that existed during the Upper Palaeozoic/Mesozoic era

Peak flow event Time interval of highest current speed

Photic zone Water depth interval with sun light intensities >1% (sufficient for photosynthesis)

Physical properties Measurable property of a physical system, here seawater

Present day sea level Present day sea surface and coastline. During the last glacial maximum, the sea level was approximately 120 m lower due to water stored on land in ice caps

Preservation potential The likelihood of an object or feature remaining recognisable over time

Pychnocline Water mass boundaries characterised by a rapid change in water density

Raster data set Geographic data stored as regular cells covering an area. Each cell has a single value assigned

Remotely operated vehicle Underwater robot connected to a ship by an umbilical for communication, control and energy supply

Re-suspend The lifting of previously deposited sediment into the water by strong bottom currents

Reworking Repeated mobilisation, transport and deposition of sediment

Ridge Generic term for an elongated elevated morphological feature

Rift An area where the Earth's crust is pulled apart and eventually new oceanic crust is formed by volcanism. When rifts deepen below sea level and become flooded, new oceans may be formed

Ripples Centimetre-sized sediment waves

Rock outcrop Rock exposed on the seabed indicative of erosive conditions

Salinity The concentration of dissolved ions in water

Sea level Elevation of the sea surface and the coast line

Seabed feature In the context of this atlas, morphologically distinct features on the seabed

Seamount Submarine banks of mostly volcanic origin that elevate more then 1,000 m above the surrounding seabed

Sediment fan Depositional feature at the mouth of canyons and channels formed by the down-canyon/channel accumulation of sediment

Sediment load Amount of sediment carried in suspension

Shelf Gently sloping seabed from the shore to the shelf break. Approximately 0–200 m water depth (see also Chapter 1.1)

Shelf break Steepening of the seabed from the shelf to the continental slope at approximately 200 m water depth. Often representing the shoreline during glacial sea level low stands

Shelf edge See shelf break

Shelf edge current Bottom current following the contour of the shelf edge

Sill Horizontal magmatic intrusion

Slide Descent/avalanche of a mass of sediment down a slope

Slope angle image Image of the seabed showing the changes in slope angle and not the bathymetry (first derivative of the bathymetry)

Slope inclination raster data Raster data set with every cell representing the average slope angle of the area of the cell

Sounding A measurement of water depth

Species diversity Number of species in a defined area

Spreading Divergence of tectonic plates with simultaneous formation of new (oceanic) crust

Taylor column Physical phenomenon of rotating fluids, in the context of this atlas: a current vortex over a seamount

Terrestrial sediment Sediments formed on land as opposed to marine sediments formed in the oceans

Thalweg Deepest path (weg) along a valley (thal)

Turbidity current Sediment laden suspension moving downslope often in channels or canyons

Upwelling Wind driven rising of cooler nutrient rich waters into the photic zone from the deeper oceans

Water mass Body of water with a common origin often characterised by it's temperatures and salinity values

Water mass boundaries Contact between water masses were density, salinity and temperature can change rapidly

WGS 84 World Geodetic System dating from 1984. WGS 84 is the reference system used by GPS

XYZ coordinates Longitude, latitude and elevation of a point on the surface of the Earth

Zooplankton Animals drifting in the water column of the ocean

Index

GPSR Compliance
The European Union's (EU) General Product Safety Regulation (GPSR) is a set of rules that requires consumer products to be safe and our obligations to ensure this.

If you have any concerns about our products, you can contact us on

ProductSafety@springernature.com

In case Publisher is established outside the EU, the EU authorized representative is:

Springer Nature Customer Service Center GmbH
Europaplatz 3
69115 Heidelberg, Germany

www.ingramcontent.com/pod-product-compliance
Ingram Content Group UK Ltd.
Pitfield, Milton Keynes, MK11 3LW, UK
UKHW050737270826
474776UK00005B/52
9789048193752